palgrave advances in
intellectual history

Palgrave Advances

Titles include:

H.G. Cocks and Matt Houlbrook (*editors*)
THE MODERN HISTORY OF SEXUALITY

Saki R. Dockrill and Geraint A. Hughes (*editors*)
COLD WAR HISTORY

Patrick Finney (*editor*)
INTERNATIONAL HISTORY

Jonathan Harris (*editor*)
BYZANTINE HISTORY

Marnie Hughes-Warrington (*editor*)
WORLD HISTORIES

Helen J. Nicholson (*editor*)
THE CRUSADES

Alec Ryrie (*editor*)
EUROPEAN REFORMATIONS

Richard Whatmore and Brian Young (*editors*)
INTELLECTUAL HISTORY

Jonathan Woolfson (*editor*)
RENAISSANCE HISTORIOGRAPHY

Forthcoming:

Jonathan Barry and Owen Davies (*editors*)
WITCHCRAFT STUDIES

Katherine O'Donnell, Leann Lane
and Mary McAuliffe (*editors*)
IRISH HISTORY

Palgrave Advances
Series Standing Order ISBN 1–4039–3512–2 (Hardback) 1–4039–3513–0 (Paperback)
(*outside North America only*)

You can receive future titles in this series as they are published by placing a standing order. Please contact your bookseller or, in the case of difficulty, write to us at the address below with your name and address, the title of the series and the ISBN quoted above.

Customer Services Department, Macmillan Distribution Ltd, Houndmills, Basingstoke, Hampshire RG21 6XS, England

palgrave advances in intellectual history

edited by
richard whatmore and brian young

macmillan

First published 2006 by
PALGRAVE MACMILLAN
Houndmills, Basingstoke, Hampshire RG21 6XS and
175 Fifth Avenue, New York, N.Y. 10010
Companies and representatives throughout the world

PALGRAVE MACMILLAN is the global academic imprint of the Palgrave
Macmillan division of St Martin's Press, LLC and of Palgrave Macmillan Ltd.
Macmillan® is a registered trademark in the United States,
United Kingdom and other countries. Palgrave is a registered
trademark in the European Union and other countries.

ISBN-13   978-1-4039-3900-5 hardback
ISBN-10   1-4039-3900-4 hardback
ISBN-13   978-1-4039-3901-2 paperback
ISBN-10   1-4039-3901-2 paperback

This book is printed on paper suitable for recycling and
made from fully managed and sustained forest sources.

A catalogue record for this book is available from the British Library.

A catalogue record for this book is available from the Library of Congress.

10 9 8 7 6 5 4 3 2 1
15 14 13 12 11 10 09 08 07 06

Printed and bound in Great Britain by
Antony Rowe Ltd, Chippenham and Eastbourne

contents

acknowledgements

The editors would like to thank the publishing and production team at Palgrave, and especially Ruth Ireland, for their help in seeing this book through to publication. They would also like to thank Katharyn Lanaro of the University of Sussex for help with editorial labours.

notes on contributors

Mishtooni Bose is Christopher Tower Official Student in Medieval Poetry in English at Christ Church, Oxford. Her research includes heresy, orthodoxy and intellectual life in the Middle Ages and her most recent publication is 'Vernacular philosophy and the making of orthodoxy in the fifteenth century', *New Medieval Literatures*, 7 (2005). She is completing a book on writing and reform from Thomas Netter to Thomas More and her next research project will be an examination of the influence of Guillaume de Deguileville on late-medieval English literature.

John W. Burrow FBA was formerly Professor of Intellectual History at the University of Sussex (1981–95). He retired as Professor of European Thought at the University of Oxford in 2000. His most recent book is *The Crisis of Reason. European Thought, 1848–1914* (Yale, 2000).

Brian Cowan holds the Canada Research Chair (Tier II) in Early Modern British History at McGill University and is the author of *The Social Life of Coffee: The Emergence of the British Coffeehouse* (Yale, 2005).

Rachel Foxley is Lecturer in Early Modern British History at the University of Reading. She is the author of a forthcoming monograph on the Levellers' conception of the political nation. She teaches and researches in the area of seventeenth-century English political thought, particularly the radicalism of the Civil War period, and traditions of political thought from classical to early modern Europe.

Lucy Hartley is Senior Lecturer in English at the University of Southampton and co-founder of the Centre for Rhetoric and Cultural Poetics. She is the author of *Physiognomy and the Meaning of Expression in Nineteenth-Century Culture* (Cambridge, 2000), and researches the development of scientific thought and the aesthetic movement in the nineteenth century.

Duncan Kelly is Senior Lecturer in Politics at the University of Sheffield. He is the author of *The State of the Political* (Oxford, 2003) and of various papers in the fields of intellectual history and the history of political thought. He is currently writing an intellectual history of the relationship between the passions, liberty and self-cultivation in modern political thought.

James Livesey is Reader in Early Modern History in the University of Sussex. His early work was on the French revolutionary tradition and its contribution to the elaboration of European democracy. At present he is working on a social history of reason in the eighteenth century. He has published *Making Democracy in the French Revolution* (Harvard, 2001) as well as articles on plough technology, in *Past and Present*, and Irish political theory, in the *Historical Journal*.

Deborah Madden is a Postdoctoral Research Fellow in the Theology Faculty at the University of Oxford. She has published several articles on the relationship between religion, medicine and culture in eighteenth-century England. Currently, she is completing a monograph on the self-styled prophet, Richard Brothers (1757–1824), as part of the 'Prophecy Project' at Oxford. This is the first book-length study to make an intellectual assessment of Brothers's theological, political and cultural significance during the romantic period.

Richard Whatmore is Reader in Intellectual History at the University of Sussex. He is the author of *Republicanism and the French Revolution* (Oxford, 2000), and a co-editor with Stefan Collini and Brian Young of *Economy, Polity, Society*, and *History, Religion, Culture: British Intellectual History, 1750–1950* (Cambridge, 2000).

Abigail Williams is Fellow and Tutor in English at St Peter's College, University of Oxford, and the author of *Poetry and the Creation of a Whig Literary Culture 1681–1714* (Oxford, 2005).

Brian Young is University Lecturer and Student and Tutor in History at Christ Church, Oxford. He is the author of *Religion and Enlightenment in Eighteenth-Century England* (Oxford, 1998), and a co-editor with Stefan Collini and Richard Whatmore of *Economy, Polity, Society*, and *History, Religion, Culture: British Intellectual History, 1750–1950* (Cambridge, 2000). He has written widely on British intellectual and religious history, and is currently completing a study of Victorian understandings of the eighteenth century.

introduction

brian young

What is intellectual history? It is the purpose of this volume to answer that question, and also to demonstrate how answering it is affected by its relations with other disciplines. The relationship between intellectual history and other disciplines matters fundamentally because intellectual history, rather more than is the case with many other branches of historical enquiry, is itself a supremely interdisciplinary enterprise. This volume demonstrates that intellectual history has pioneered and will continue to promote much interdisciplinary activity, both among historians themselves, and also in the practice of allied disciplines, especially in the congruent field of literary studies. It ought to come as no surprise, therefore, that several of the contributors to the present volume are scholars of English literature. As the chapter by Abigail Williams demonstrates, however, literary history has its own distinctive characteristics, which are allied to, but are not identical with, those attached to intellectual history.

This volume is centrally concerned with the complementary aspects of those disciplines that most clearly and fruitfully employ the techniques associated with intellectual history rather than with justifying the ready reduction of all such enterprises to a tendentiously, all-encompassing, pseudo-Hegelian form of intellectual history. As John Burrow warns in his scene-setting chapter, it is tempting to adduce all human thought to the province of the intellectual historian, but this, like most temptations, is a temptation to be resisted, however reluctantly. Disciplines exist precisely in order to instil intellectual rigour; all reputable interdisciplinary activity has, therefore, to be conducted with exemplary finesse. The chapters in this volume are dedicated to promoting just such attention to nuance and an allied attentiveness to the distinctiveness discernible when one is working within differing intellectual and disciplinary conditions.

As John Burrow intimates in his chapter, intellectual history might most readily be defined with reference to a series of volumes published by Cambridge University Press under the title of 'Ideas in Context'. That is, intellectual history is concerned with understanding how ideas originate and evolve in specific historical contexts; it is also concerned with tracing their histories within the broader histories of the societies and cultures which they have helped to shape, and which have also shaped them. There are, of course, hermeneutical difficulties with the composition of such histories: how far, for example, did the revolutionary political thought of John Locke serve to enact such later events as the American War of Independence, and how far was the subsequent history of those ideas shaped by the assumption, not to say presumption, that they had somehow been a contributing cause of those events which we identify as the American War of Independence? Similarly, modern apologists for Jean-Jacques Rousseau insist that his models of a transformed political society were specifically *not* designed for the revolutionary transformation of France; how, and why, though, were they subsequently adopted and shaped by such ideologues as Saint-Just and Robespierre to precisely such allegedly inappropriate ends?[1] At what identifiable point do textual ideas, embodied in treatises, become doctrinal ingredients in political society? These and allied difficulties plainly complicate the resolution explicitly made by the locution of 'ideas in context', but complication is not the same as refutation, and intel-lectual (and cultural) historians have refined their practice in ways that attend to precisely such a collection of difficulties. Robert Darnton, for example, has shown how one disciple of Rousseau read his hero's books and sought to bring up his children according to what he took to be Rousseau's ideals; unfortunately (or fortunately?), we do not know what the eventual results of this educational experiment were.[2] As the chapter by Brian Cowan contends, a richly suggestive cultural history of ideas is developing that can begin to address such problems, a mode of historical understanding in which Darnton, alongside such scholars as Roger Chartier, can be seen as a pioneer.

'Ideas' necessarily take many shapes, not all of which, as Burrow argues, are readily identifiable with the forms created and enforced by modern disciplines. As Burrow also shows in this connection, the world that we dismiss as that of 'magic' flowed into what we all too easily identify as science, but it did so neither simply nor definitively. Perhaps a cultural history of current 'New Age' beliefs might account for the continuing presence of such phenomena, but few, if any, modern scientists would be happy to make any accommodation for them in their own worldviews. As the chapter in this collection by Deborah Madden demonstrates, the history of medicine incorporates a decidedly critical analysis of the cultural contexts of a medical universe promoted by suitably qualified practitioners in an increasingly professionalised field of human endeavour. The history of medicine, as instanced in the work of Michel Foucault, whose work Madden analyses, is very much a history of a series of disciplinary interventions imposed by a powerful cadre of modernising specialists. The chapter by Jim Livesey likewise draws our attention to the professionalising disciplinary ethos in the history of science, of which he is properly and productively critical, discerning in its stead the continuing need for the cultivation of a culturally critical history of science. The chapters by Madden and Livesey demonstrate just how attractively complex a disciplinary history now has to look in the wake of the contributions of such seminal thinkers as Foucault.

In proceeding in this disciplinary manner, two missing elements in this volume have to be explained. The first is the absence of a chapter concerning 'philosophy and intellectual history'. The study of philosophy, after all, centrally involves study of what previous philosophers have written, and it often does so historically. There is, however, such a phenomenon as philosophers' history, which is a history very much concerned with elucidating what modern philosophers find to be of continuing interest in the work of dead philosophers; this can produce illuminating and suggestive work, but, at its worst, it can also turn into a somewhat one-sided dialogue. For three fairly recent and culturally significant examples of how a distinguished philosopher can prove a poor intellectual historian, one might instance the short and deeply unsatisfactory studies of Hume, Voltaire and Paine written by the late A. J. Ayer.[3] Excellent historical practice in the field of the history of philosophy, on the other hand, can be found in the newly established series 'Studies in Early Modern Philosophy', published by Oxford University Press, and in the work of such established scholars as Knud Haakonssen and such younger contributors to the field as James Harris.[4] Philosophy is, necessarily, a deeply reflexive practice, and it has naturally played a major role in the evolution of intellectual history as a discipline. The major practitioner of such enquiry in England was the late Sir Isaiah Berlin, whose contribution is discussed in the present volume in the chapters by Brian Young and Duncan Kelly. One can also point to the plea for the world of thought in the practice of historical enquiry made by another philosopher with an interest in history, R. G. Collingwood, who famously argued in his influential study *The Idea of History* that 'All history is the history of thought.'[5] In seeking to develop that claim, however, we are in some danger of becoming the megalomaniac historian against whose potentially disastrous example Burrow rightly warns us.

Where, in short, does the practice of philosophy take over from that of intellectual history? This is a difficult question which cannot be resolved with any degree of precision. It is, however, one to consider seriously when reading the contributions to this volume. Is intellectual history, even more than history *per se*, 'philosophy teaching by example'? If so, exactly what sort of history is it? Quentin Skinner has recently argued for a politically aware practice of intellectual history, eschewing in the process what he calls antiquarianism in favour of a relevant and sustaining historical practice which deepens contemporary understandings of political problems.[6] Is this modern form of historical humanism the sort of marriage between intellectual history and philosophy that would bring intellectual history centrally into modern life? Just how this variety of intellectual history is beginning to look can be seen in the work of such disparate philosophers as Alasdair Macintyre, Susan James, Charles Taylor and the late Bernard Williams; it is a rich and provocative literature that vitally links intellectual history with contemporary philosophy.[7]

The other field that is missing from the present conspectus of chapters and themes is music. Musicology has defiantly come of age intellectually, and its attention to the disciplinary problems attendant on its practice has proved exemplary. To give but one instance, the desire to play pre-modern music 'authentically' has raised a plethora of problems of a decidedly historical kind, and these have their parallels in the pursuit of a hermeneutically sophisticated intellectual history.[8] The work of John Butt in particular has taken these problems to new heights of articulation, and intellectual historians would do well to consider his thoughts on this deeply important matter.[9] As Tim Blanning has recently demonstrated to great effect, the history of music is a vital part of the cultural history of the modern West; it is a sign of the professional philistinism of much historical enquiry, however, that classical music continues to be ignored in assessments of modern culture, despite the exciting work that musicologists have recently undertaken when thinking about such composers as Alban Berg and Benjamin Britten in their historical and intellectual contexts.[10] After all, an intellectual history of modern Europe that failed to register an interest in Richard Wagner would be a somewhat barren exercise. There are, of course, technical matters which make such discussions difficult for lay readers, but, as with philosophy, the history of music makes interpretative demands of non-professionals which musicologists are beginning to address. It is well to consider, albeit momentarily and fleetingly, what such reflection might contribute to the future practice of intellectual history.

Lucy Hartley shows in her chapter how the neighbouring field of art history has long related to the territory more usually surveyed by intellectual historians. The history of art emerged as a subject at much the same time and in similar ways to intellectual history, and it is remarkable how many of its mid-twentieth-century practitioners were refugees from racial and political persecution, a humbling demographic component it also shares with intellectual history. As the chapters by Hartley, Young and Kelly demonstrate, the intellectual diaspora that followed the rise of Fascism and Nazism contributed in no small way to a new degree of intensity and richness in British and North American scholarship. Peter Gay's moving memoir, *My Germany*, for example, charts how this leading historian of Enlightenment Europe and Weimar culture made his way from a secular childhood in Berlin into a new American life, from which, influenced by his reading of another Jewish exile, Sigmund Freud, he was able to analyse the immediate prehistory of the sophisticated European culture that all too quickly gave way to racial hatred and the unspeakable horrors of genocide.[11] His is an experience shared with John Clive, whose work on Victorian intellectual history is exemplary for its literary sensitivity and pervasive love of culture; it is as if the horrors of his early experience had deepened his appreciation of the high culture of a past that was largely (but not entirely) unblemished by the atrocities which followed.[12] We still have much to learn from this heroic generation of scholar refugees. Some refugee scholars, notably Isaiah Berlin, instinctively became historians of exile, and it was the example of Berlin's work in this connection that led to Tom Stoppard's stunning trilogy of plays about nineteenth-century Russian intellectual exiles, *The Coast of Utopia*, replete with a telling scene involving Ivan Turgenev in conversation with a doctor in Victorian Ventnor.[13]

The politics of intellectual history is the subject of Duncan Kelly's chapter, which surveys the discipline's fortunes during the deeply troubled twentieth century. In the course of his survey, Kelly shows that intellectual historians could and did serve a persecuting tyranny, as the troubling example of Carl Schmitt testifies. How such men managed to justify such a betrayal of the life of the mind let alone a basic sense of humanity remains mysterious, but for some sense of the peculiar horrors attendant on it one can firmly recommend a reading of a powerful poem on the theme by Geoffrey Hill, 'Ovid in the Third Reich'. As well as helping to shape the experience of the twentieth century, intellectual history was shaped by it, and that has negative as well as positive implications for those assessing the consequent fortunes of the discipline.

One of the more positive intellectual and cultural products of the twentieth century was feminism, and Rachel Foxley demonstrates in her chapter how feminist scholarship has begun to deepen our appreciation of the intellectual history of Europe and North America. Taking a number of case studies, Foxley makes an eloquent argument regarding the consequences for the history of political thought in particular of applying some of the categories and questions that feminist scholars have importantly begun to ask of the past. Her chapter details a significant advance in the practice and principles of intellectual history from a committed perspective. Commitment of an allied kind marks Richard Whatmore's analysis, in his chapter, of the revival of republicanism as a dominant theme in the recent history of political thought, the register that has enjoyed a peculiar dominance in the practice of intellectual history in Britain. It is worth asking why political thought has long enjoyed such a privileged position in the history and practice of intellectual history in Britain.[14] Is it related to the importance of constitutional history in Britain, the sense that the strength and continuity of its political institutions, particularly when compared with those of other European countries, guarantees its historians a fascination with its own past? The fortunes of Thomas Hobbes, James Harrington and John Locke in the history of the history of political thought might provide an interesting lesson in themselves. It is certainly a significant fact that the revival of republicanism as a topic for research seems to indicate some unease with the current constitutional settlement in Britain, a debate which surely merits much more public discussion than present socio-political conditions would seem to allow.

The ventilation of the issues raised by republicanism relates very centrally to perceptions of the public intellectual, still a relatively unfamiliar concept in Britain. The public intellectual is a much more familiar manifestation of what Pierre Bourdieu called '*Homo academicus*' in mainland Europe than it ever was in modern Britain.[15] As Mishtooni Bose's chapter demonstrates, the ideal of the public intellectual has strongly affected recent conceptions of the 'intellectual' in the Middle Ages. Allied to this is the centrality of religion in medieval studies, perhaps the only genuinely interdisciplinary field at present practised by historians and other scholars. The presence of Bose's chapter in this volume precluded the need for a separate chapter on theology and intellectual history, so deeply imbricated are the two in medieval studies. The example of medieval studies is indeed an optimistic one for interdisciplinary scholarship, not least in the form of intellectual history, whose procedures are especially well suited to the sort of nuanced, precise and attentive scholarship that it is dedicated to achieving. Bose's concluding endorsement of an embedded, materialist intellectual history (rather like Cowan's culturally inflected form of intellectual history), provides one understanding of how 'Ideas in Context' must continue to be at the core of intellectual history.

Medievalists have also been amongst the foremost intellectual historians seeking to relate positively to postmodernism in history. Whilst such modern intellectual historians as Joyce Appleby, Lynn Hunt and Margaret Jacob have sought to resist the advances of postmodernism in history, students of late antiquity such as Elizabeth A. Clark and medievalists such as Gabrielle Spiegel, whom Bose also takes into account, have called on historians to be more critically receptive to postmodernism as an approach to the past.[16] In so far as intellectual history is supremely concerned with texts, and to understanding 'texts' as taking a variety of forms – literary, cultural, ritualistic – then it has to be attentive to postmodern calls for acute self-consciousness on the part of ourselves as interpreters of such texts. Without explicitly adopting such language or approaches, it is hoped that the chapters in this volume show a level of receptivity to these developments that bodes well for future advances in intellectual history.

notes

1. For a fascinating discussion of these issues, see Carol Blum, *Rousseau and the Republic of Virtue: the language of politics in the French Revolution* (Ithaca, NY, 1986).
2. Robert Darnton, 'Readers Respond to Rousseau: the fabrication of Romantic sensitivity' in *The Great Cat Massacre and Other Episodes in French cultural history* (London, 1984), pp. 209–49.
3. A. J. Ayer, *Hume* (Oxford, 1980); *Voltaire* (London, 1988); *Thomas Paine* (London, 1989).

4. Knud Haakonssen, *Natural Law and Moral Philosophy: from Grotius to the Scottish Enlightenment* (Cambridge, 1996); James Harris, *Of Liberty and Necessity: the free will debate in eighteenth-century Britain* (Oxford, 2005).

5. R. G. Collingwood, *The Idea of History* (Oxford, 1946), p. 115.

6. Quentin Skinner, *Liberty before Liberalism* (Cambridge, 1998).

7. Alasdair MacIntyre, *After Virtue: a study in moral theory* (London, 1981), and *Whose Justice? Which Rationality?* (London, 1988); Susan James, *Passion and Action: the emotions in seventeenth-century philosophy* (Oxford, 1997); Charles Taylor, *Sources of the Self: the making of modern identity* (Cambridge, 1989); Bernard Williams, *Truth and Truthfulness: an essay in genealogy* (Princeton, 2002). Alasdair MacIntyre and Charles Taylor were also contributors to an important collection of essays by philosophers and historians which can still be consulted with profit: Richard Rorty, J. B. Schneewind and Quentin Skinner, eds, *Philosophy in History: essays on the historiography of philosophy* (Cambridge, 1984).

8. For an argument which seeks to integrate these parallel histories in a preliminary way, see Brian Young, 'The Tyranny of the Definite Article: some thoughts on the art of intellectual history', *History of European Ideas* 28 (2002), 101–17.

9. John Butt, *Playing with History: the historical approach to musical performance* (Cambridge, 2002).

10. T. C. W. Blanning, *The Culture of Power and the Power of Culture, 1660–1789* (Oxford, 2002); Anthony Pople, ed., *The Cambridge Companion to Berg* (Cambridge, 1997); Mervyn Cooke, ed., *The Cambridge Companion to Benjamin Britten* (Cambridge, 1999).

11. Peter Gay, *My German Question: growing up in Nazi Berlin* (New Haven., Conn., 1998); *The Enlightenment: an interpretation* (2 vols, New York, 1966–69); *Freud, Jews, and other Germans: masters and victims in modernist culture* (Oxford, 1978); *Weimar Culture: the outsider as insider* (New York, 1968); *The Bourgeois Experience: from Victoria to Freud* (5 vols, New York, 1985–98).

12. John Clive, *Scotch Reviewers: the* Edinburgh Review, *1802–1815* (London, 1957); *Macaulay: the shaping of the historian* (New York, 1978); *Not by Fact Alone: essays on the writing and reading of history* (London, 1989). For a moving appreciation, see the memoir in Susan Pedersen and Peter Mandler, eds, *After the Victorians: essays in memory of John Clive* (London, 1994).

13. See the acknowledgements to Tom Stoppard, *The Coast of Utopia* (3 vols, London, 2002), which declare his indebtedness to the work of Berlin and also to E. H. Carr's *The Romantic Exiles*.

14. On the remarkable and powerful longevity of the history of political thought within the practice of intellectual history in Britain, see for its inception, Mark Goldie 'J. N. Figgis and the History of Political Thought in Cambridge' in Richard Mason, ed., *Cambridge Minds* (Cambridge, 1994), pp. 177–92, and, for its current standing, Anthony Tuck, 'History of Political Thought' in Peter Burke, ed., *New Perspectives on Historical Writing* (2nd edn, Cambridge, 2001), pp. 218–32, and Annabel Brett, 'What is Intellectual History Now?' in David Cannadine, ed., *What is History Now?* (Basingstoke, 2002), pp. 113–31.

15. Pierre Bourdieu, *Homo Academicus*, trans. Peter Collier (Cambridge, 1988).

16. Joyce Appleby, Lynn Hunt and Margaret Jacob, *Telling the Truth About History* (London, 1994); Gabrielle Spiegel, *The Past as Text: the theory and practice of medieval historiography* (Baltimore, Md, 1997); Elizabeth A. Clark, *History, Theory, Text: historians and the linguistic turn* (Cambridge, Mass., 2004).

1

intellectual history in english academic life: reflections on a revolution

john w. burrow

Just under 20 years ago I gave a lecture in the University of London Senate House on the state of Intellectual History in England in my academic lifetime. It consisted of two themes. The first was the lack of recognition, as it seemed to me, accorded to the subject. The second was a number of caveats about the forms that such recognition, in a better world, might take.

Of course we all tend to think our subjects undervalued and under-resourced, but the improvement of the academic standing of Intellectual History or the History of Ideas (I use the terms interchangeably here) has been dramatic and scarcely precludes the use of the term 'revolution'. I may, of course, have been unduly surly in 1987. It depends on one's expectations, though I can cite a high authority who did not, we know, feel personally undervalued, in support. Isaiah Berlin, who was present, wrote to me afterwards saying he recognised the situations I had described. The remarkable improvement in status and recognitions achieved in the past two decades also seem to me indisputable. I can mention just a few symptoms, beginning with the existence now of two reputable English journals devoted to the subject, apart from the more specialised *History of Political Thought*. The high standing of the latter subject in Cambridge, in particular, is well known and the Regius Chair of History is now held by the scholar who has done more than anyone to promote it there. The immense number of valuable monographs published by Cambridge University Press in its Ideas in Context series is another impressive symptom. The establishment in Oxford in 1994 of a chair of European Thought seemed a portent, though for extraneous reasons it has proved abortive. Majors in Intellectual History have been established since one was founded in the University of Sussex in 1969, as have various MAs and M.Phils. Impressionistically it seems to me that there is more interest in the subject among postgraduate students than there has ever been, including, at last, an awareness that it is not co-extensive with the history of philosophy.

My 1987 lecture, therefore, is in some respects happily out of date, though an historical retrospect of how things were is perhaps not without interest. The caveats, however, I see no reason to withdraw. On the contrary, the more self-confident we become, the more they seem likely to be relevant.

i

I began the lecture with a quotation from A. E. Houseman in his 1892 Inaugural lecture as a Professor of Latin at University College London. As a quip it was rather successful, as Houseman's quips usually were, and I warned the audience that I proposed to spend the next hour labouring the point of a century-old academic joke. Their patience was admirable.

'Richard Bentley', Houseman told his audience, 'was born in the year 1662 and he brought with him into the world, like most men born near that date, a prosaic mind.'[1] It was something to have had the history of ideas recognised by Houseman, even if somewhat obliquely and parenthetically.

What we might perhaps have induced Houseman himself to say, more directly and less ironically, about the subject of the history of ideas I admit I cannot say, but the question is not quite an absurd, in the sense of anachronistic, one. In fact it occurs to me that the teasing reference to the Bentleyan *Zeitgeist* could have been prompted by a work which had appeared a few years earlier, which had better claims than any other I know to be called the first really extensive and detailed work of English intellectual history; I mean Leslie Stephen's *English Thought in the Eighteenth Century*, published in 1876. Stephen, his first biographer Maitland tells us, 'was ... impressed by the interdependence of all kinds of thought – theological and metaphysical, political and poetical, scientific and fanciful'. It is indeed, I think, the sense of that interdependence which makes Stephen's book really a history of past thought, compared with the more jejune exercises we find in what might perhaps be called the prehistory of the history of ideas in England, earlier in the last century: works such as, for example, James Mackintosh's *Progress of Ethical Philosophy* or George Henry Lewes's *Biographical History of Philosophy*. As it proceeds, I have to admit, Maitland's account of Stephen's views becomes, from my point of view, less encouraging: 'Later on I have heard him maintain that philosophical thought and imaginative literature can have no history, since they are but a sort of by-product of social evolution, or, as he once put it, "the noise that the wheels make as they go round".'[2] It is a view we are accustomed to hear more ponderously expressed in terms of 'historical materialism', rather than of 'social evolution', but it is, of course, as Stephen's case reminds us, and as we shall see later, by no means confined to its professed devotees.

But why bother to raise the question of the bare existence of the history of ideas? Surely its existence is a matter of common knowledge and common sense. We know that we do not think in important ways exactly as our immediate forebears did, nor they like theirs; ergo it exists. Indeed yes. But also, alas no. For we have to speak not only of common sense but of academic consensus, which is not invariably the same thing, not of what every schoolboy knows but of what his syllabus requires him to know about. In speaking of the existence of the subject of my lecture as debatable I spoke not, indeed, sensibly, but I spoke academically. I was, of course, exploiting that ambiguity by which we may sometimes speak of an academic subject both as a form of knowledge and as what it knows, and of 'history' both as the practice of historians and as the matter of the past. Not that I at all wanted to deny that in the history of ideas much distinguished practice went on, though not all of it by professed historians. But my subject was not the distinction between the substance of the intellectual past and the practice of scholars in dealing with it but with a remoter one, between the practice of a subject and the academic self-consciousness which demands recognition of it *as* a subject, a form of scholarly practice.

I think it is true that the consensus of professional historians has not always readily allowed such recognition to the history of ideas, at least as their kind of business, nor, I dare say, do all those who, to my eye, practice some form of it, always claim or perhaps see any virtue in such recognition. It is in that sense that I speak of its existence as debatable. Of course, a precondition of recognition is the academic or scholarly self-consciousness that claims recognition, and here, in fact, I want to enter a caveat. I shall argue for such self-consciousness on behalf of the history of ideas, but I do not think of it, especially when it takes the form of a demand for an overarching theory or a distinctive scholarly vocabulary for the subject, as by any means an unmixed blessing. Such demands, in the case of the history of ideas, I shall argue later, may be not so much 'premature' – always a tempting evasion – as actually misconceived or in some respects mischievous. History is notoriously a relatively piecemeal and untheoretical discipline – rightly and inevitably, I am sure, given what historians attempt to do – and I see no reason for thinking of the history of ideas differently in this respect. So, in endorsing awareness of the alienness, the otherness, of the intellectual life of the past, under the name of the history of ideas, or intellectual history, I am anxious not to be suspected of saying more than I mean or of forgetting that the solemnities of academic self-definition have often more to do with public relations than with practice, and that academic labels are better thought of as flags of convenience than as names of essences.

But we need initially, of course, to make the case for self-consciousness and recognition, and to suggest what forms that awareness may desirably take.

Perhaps it will be as well to begin somewhat crudely with a definition. The history of ideas or, intellectual history attends, as I take it, to the reflective communal life of human beings in the past; to their assumptions, arguments, enquiries, ruminations about the world and themselves, their past and their future and their relations to each other, and the various vocabularies and rhetorics in which they conducted these. This sounds an ambitious programme, but less so, after all, than simply 'History'.

Incidentally, I myself mildly prefer 'intellectual history' to the perhaps more familiar 'history of ideas', as registering, by analogy with 'political' or 'economic' history, an attention to forms of human activity rather than to some historical encounter of abstract categories. In this chapter, however, I shall use the two phrases interchangeably. Why the activity so described should be thought more obscure, more puzzling or more perverse, as a way of spending one's time in the library and the classroom, than looking at how people ran their public affairs, got their livings or pursued power or wealth is not self-evident. But it would be naive to pretend that it has not been felt to be so: that a label saying 'intellectual history' could be passed off as casually as if the adjective were 'political' or 'economic' or even, more mysteriously, 'social'. To take only one, admittedly rather trivial, example of the relative English inhospitality to the concept of intellectual history there are, of course, the resonances of the word 'intellectual' itself, which sometimes used to be an embarrassment to my pupils at Sussex whose choice of subject designated them as 'Intellectual Historians'. I consoled them by pointing out that it is only an adjectival quirk of the language which makes it seem more obligatory on intellectual historians to be intellectual than on economic historians to be thrifty or social historians to be gregarious, and I also asked them to consider how much worse off they would be if their metier obliged them to call themselves criminal lawyers. Some, nevertheless, give me the impression that in England the balance of advantage between being intellectual and being criminal remained debatable.

ii

This relative inhospitality of English culture to the idea – and I think it is the idea as well as the phrase – of intellectual history would itself make an interesting subject for an essay in English intellectual history. I stress 'English' because on the Continent matters have been different, with the impressive German traditions of *Kulturgeschichte*, successfully transplanted in England in the Warburg Institute, and where also the history of 'philosophy' has tended to be interpreted more widely than in English philosophy departments. In the United States too, the history of ideas took root earlier. We could narrow the question of the relative English coldness to it, I think, to more manageable proportions by looking at the ways the historical profession, in particular, developed and the influences under which it did so. One way, in which the concerns of English historians have been conditioned, is by the early emergence of England as a strong and unified state, and by the continuity of its institutions. The contrasts with Germany, particularist, federal, unified as a nation-state only in 1871, with Italy and her somewhat similar history, and with the rupture in French history made by the first French Revolution, are very marked. When European historiography developed into a profession in the nineteenth century, under the influence of powerful nationalist sentiment and urged on in part by the quest for national identity, the English seemed to have an identity already apparent, which was overt, practical, political and institutional, rather than primarily cultural or intellectual, and in so far as the latter was attended to it was academically focused as 'English', that is English literature, rather than history. It was otherwise in Germany, which could claim a culture long before it was a state; German identity was necessarily initially the identity not primarily of institutions, except at the most local level, but of '*Geist*' or '*Kultur*', the subject of the philosophic or cultural historian, and was early acclaimed as such. In France, at least for the good anti-clerical republican, the continuity and identity of modern France lay not in a constitution or polity, which had suffered many vicissitudes since 1789, but in a secularist and revolutionary republican ideological tradition which traced its roots to the eighteenth-century Enlightenment.

In England, the disposition to assume that the stuff of history was how institutions worked or what politicians had done – a predisposition natural in any case to an historical profession inclined to adopt research in public archives as its badge of professional distinctiveness and respectability – had no such uncomfortable historical circumstances to accommodate. Fifty years ago, when I began to read history as an undergraduate, much conspired to reinforce the confidence of the English political and administrative historian. It was, after all, an era then hailed, in one of the less fortunate historical guesses of the century, as 'the End of Ideology'. For the aspiring young historian an often tacit but powerfully influential metaphysics distinguished sharply between the real and weighty and the unreal or vapid. There were real historical questions – usually about how the apparatus of government worked or who, in running it, fixed what, or whom, and how. And there were unreal or 'irrelevant' questions, chiefly what anyone, in a general way, thought about it all, or indeed about anything, and the more general and coherent the thoughts the more the occult quality of reality was deemed to have leaked from them. In implying, as I may perhaps have done, that I found this metaphysics uncongenial, as I now find it dogmatic and unpersuasive, I do not, of course, want to reverse its terms – to imply that what was thought important is not, indeed, important, but only to suggest that it was unjustifiably restrictive in its exclusions. And of course in so describing it I simplify and caricature and to at least some of my seniors I appear to do injustice. Among my own teachers, Duncan Forbes, Peter Laslett and Noel Annan were very honourable exceptions.

Well it may be said, much changed even in the ensuing 30 years or so. Fewer historians explicitly subscribed to the anti-intellectualist rigours of the metaphysics I have spoken of, then so powerfully exemplified and propounded by a remarkable scholar and ideologue, Sir Lewis Namier. Yet I would hesitate to claim that the history of ideas benefited as much from the change as one might have expected. For the one presupposition has been joined or partly supplanted by another, but neither was one I was anxious to embrace. The older one was that an historical subject derives weight and reality from the concept of power, explanation of whose exercise was at the centre of the historian's trade. It made room for ideas, if at all, only on terms which made their irrelevance the more starkly apparent. The one kind of intellectual history one could normally study as part of a history syllabus was something called the history of political thought or theory; indeed, the power of habit is such that I teach it. But it sometimes struck me, I think, even at 21, as an odd restriction. The seventeenth century, it seemed, contained Hobbes and Locke but not Bacon, Descartes or Newton. The eighteenth century held Montesquieu and Rousseau, but not – unless one was lucky enough, as I did, to do a Special Subject on the Enlightenment – Voltaire or Adam Smith, who belonged respectively to French and Economics. Moreover, of course, one read Locke on government but not Locke on human understanding, Rousseau on the Social Contract but not on religion or education. Given that History was largely past politics, was it perhaps that we were learning about the relation of theory to practice? Perhaps Whig politicians had sat at the feet of Locke, authoritarian ones at those of Hobbes, Revolutionists presumably consulted their Rousseau, administrators their Bentham. But since in many cases it seemed that this had not been so, and since the rugged simplicity of the implied relationship between language and culture and political action more or less ensured that it would not have been so, the emancipation of the historian from attention to past thoughts seemed actually endorsed by the one form of attention given to them, and the visits to Hobbes, Locke and Company often seemed a matter of courtesy, a perfunctory leaving of visiting cards with no intention of closer acquaintances.

That was the old wisdom. The newer historical wisdom typically accommodated ideas on rather different terms. Something like a democratic revolution sometimes seemed to have taken place: ideas are important to the historian in proportion to the number or the height of their social position. Again it is not a criticism of a form of history to say that it is not intellectual history as I understand it; myself I merely wanted to claim that it is not a compliment to it either. To each kind of historical enquiry is its own appropriate sources and criteria of relevance, and they are not in competition just as ideally they should not be considered in isolation from each other. And so – to be still more platitudinous – not every question about ideas held in the past seemed best addressed by asking what the population at large thought about them more by reference to the increasingly long and complex annals of the poor. The Renaissance was not made the subject of a referendum. Fear of intellectual snobbery can be as limiting to the historical imagination as intellectual snobbery itself, and, to quote the words of two historians for both of whom I have a great respect, 'it is not only the poor and inarticulate who suffer from "the enormous condescension of posterity"'.[3]

But, surely, it may be said, speaking only of history and historians was too parochial. Surely some of the most distinguished contributions to our understanding of the intellectual life of the past came in the mid-twentieth century not from professed historians, but from philosophers, literary critics, theologians, scientists. And is that not both what we should expect and as it should be? The student who is likely to have the keenest response to, and even understanding of, some past intellectual activity, may be presumed to be one who has himself or herself felt the pressure of the same kind of question and knows what it is to attempt to give an answer. Hence the metaphysician or epistemologist will be our best guide to past metaphysical ideas, the biologist to the history of biology, and so on; indeed, the technicality and durability of some intellectual pursuits may make this virtually a tautology; he who writes the history of logic must himself be a logician. And so we arrive at a conception of the academic division of labour, as applied to the intellectual life of the past which conveniently coincides with corresponding divisions in the present, leaving each academic concern bottling its history on its own premises or premisses. All of what I have just said seems to be true until we try to generalise it as I have just done. That is, it seems to be true as far as it goes, but neither exhaustive nor devoid of certain characteristic dangers when its results are considered as history. I do not wish to deny either that a lot of what I have described went on and perhaps always will. But it cannot plausibly or safely be regarded as exhaustive of what can and should be done in the name of the history of ideas. This is not said in the spirit of the closed shop, implying that to write illuminatingly about the intellectual life of the past one needs a membership card from the Amalgamated Society of Intellectual Historians, Cultural Historians, Historians of Ideas and Allied Trades, or, more practically, that I think that it would be desirable that all practitioners of the history of ideas should be gathered together in departments of that name. There might be gains in this, but there would be losses, in terms of intellectual introversion and the loss of certain kinds of stimulus to, and ability in, the interpretation of the past, even if it were practicable.

Nevertheless, there was and are obvious objections to the present academic division of labour as the basis for our investigation of the intellectual life of the past. It might seem academically convenient. The appropriate departments, supposedly, already exist, and all that is needed is for them to display the necessary tolerance towards those of their members of an antiquarian turn of mind. The position can even be regularised with specific appointments, so that the department acquires a kind of trial bard, embodying the folk-memory, and recording *ex officio*, the great deeds of the heroes of old. I nearly, I think, began my own career in such a role. But once one has accepted such a picture, doubts and awkward questions begin to intrude. What of subjects which have emerged only recently into some kind of academic recognition, like sociology or psychology? Is it necessarily or even typically the case that the modern social scientist is our best guide to the 'social theories' of the past – and what, as one presses further back, is to count as such? And if there are subjects which lack a determinate ancestry, what of those which seem to lack an identifiable academic progeny? Must we leave the history of the influence of astrology, that subject so profoundly important to some of the most interesting minds of the Renaissance, to be written for us by gypsy ladies in tents? What, again, *counts* as continuity? Theology, philosophy and jurisprudence once divided up between them far more of the intellectual world than they do now. To whom, then, does that rich past now belong?

iii

It may seem only as we press back further into the past that the questions become awkward or absurd, but that means only that the more recent past may be more subtly and less obviously elusive if approached in this way. Disciplines are unstable through time, and confidently to superimpose the academic map of the present – itself, in places, a contentious matter – over the often very different ones current in the past is already to have taken a large step towards systematic historical misunderstanding. And if the academic map is not stable, it is also not exhaustive of everything we may think of as constituting the intellectual life of a society; a survey of modern intellectual concerns and debates which confined itself to rehearsing, *seriatim*, the state of play in the various academic disciplines, would, I suggest, strike us as both restrictive and over-rigid. It is for similar reasons that we employ, as part of the larger coinage of history, such familiar terms as 'Renaissance', 'Enlightenment' and 'Romanticism'. However much these need further analysis, as they clearly do, they do not necessarily need breaking down into academic disciplines, even those of the past, and certainly not into those of the present. It is a futile and absurd exercise to try to decide to which of our disciplines a work like Rousseau's *Discourse on the Arts and Sciences* or Bernard de Mandeville's *The Fable of the Bees* 'belongs'. It does not now belong at all, and if we choose for our own purposes to incorporate it, as we may, that in no way affects its original meaning and status. For better understanding of it we need as a mere starting point, to transpose the question into concepts of genre appropriate to the time in which it was written.

Moreover, it is not only a question of inappropriate categories. Sympathies and intellectual habits conditioned in the analogous modern discipline, supposing there to be one, may not place historical understanding high among their concerns, nor is there any reason why they should do so. It is only an obligation to do history if history is what one purports to be doing, or more obliquely, if one purports to be rendering what some past text or utterance 'meant'. And so, without denying to the modern academic the right to find whatever stimulus may be available in past texts we do, for example, need to question the credentials, as history, of one well-known genre, once particularly prevalent in the social sciences though now less popular than it used to be: the modern sociologist's or anthropologist's or economist's headmaster's report on the supposed but all too often deficient attempt of Montesquieu or Rousseau or Adam Smith to 'contribute' – a favourite word – to the development of their discipline: 'could do better', ' needs to pay more attention to empirical data', 'deplorably neglectful of recent developments', and even 'does not appear to be trying'.

Such 'histories' are, of course, examples in intellectual history of what Sir Herbert Butterfield classically christened 'the Whig interpretation of history', and it was highly characteristic of them to proceed by assuming the existence of a canon of great texts which constitute the 'history' – 'sequence' would often be a better word – of the discipline. The central characteristic of Whig history in the sense we derive from Butterfield is not merely anachronistic judgement but a particular kind of selectivity: a selection from the life of the past in terms of a notion of significance which is derived not from the conversation of the past but from what appears to be pregnant or prophetic for the future, and most specifically, of ourselves. Whiggishness of this kind is inimical to intellectual history because the latter, if it does anything, attends in the first instance precisely to the conversations of the past, to what was then found significant. Yet it is natural to many people to assume that the ideas to be particularly attended to are those which seem embryonically ours, while the archaisms which surround them, the magic, the millenarianism, the metaphysics or whatever our sense of modernity and its anticipations excludes, can be passed by as irrelevant, embarrassing, even somehow aberrant. But even apart from the claims of history, as a discipline, to be concerned with what was significant in the past as well as with what may seem significant as its potentialities, what the headmaster's report systematically ignores by its assumption of familiarity is historical distance. It is that among the ways in which texts or utterances may be difficult, intractable or opaque to us, in addition to being encoded, occult, ambiguous, fragmentary, cryptic, corrupt, ironic, and so forth, is that they can be obscure because of their historicity – obscure or misleading, that is, in ways they would not have been to their first readers, because they were written in and for a world which in various crucial ways is not ours, employing concepts and resting on conventions and assumptions which are either manifestly not our own or which, while they may seem similar, may in fact have been subtly different. And to understand past utterances better by bringing to the light their historicity we need characteristically, as the work of Quentin Skinner in particular has consistently claimed, to go beyond the canon of great texts to look at the wider discourses or intellectual contexts in which their guiding conventions and assumptions were embedded, at the web of meanings which gave the particular utterance or texts their contemporary significance.

At this point there may perhaps, understandably, be some inclination to say that all one is doing is talking about the difference between intelligent, sensitive, erudite reading and crass, self-indulgent, ignorant reading, and trying to annex the whole former for the supposed trade of the history of ideas. With this I have some sympathy. I am even prepared to be flattered by it. It would be pleasant to think of the kind of things one does as co-extensive with all sound and sensitive scholarship, but one should, I think, reserve the thought for one's megalomaniac moments. Because there is a further distinction to be made.

It concerns one's reasons for doing it, and it can best be explained by dwelling for a moment on the implications of a now old-fashioned word, once used to draw attention to the intellectual contexts of past texts. It is the word 'background', once freely resorted to by, in particular, literary critics, for referring to the kinds of thing the conscientious critic needed to be aware of but also probably ought not to spend too much time on, and it included, among other things, the intellectual history of Europe. The phrase was chiefly popularised, I suppose, by Professor Basil Willey, who in the mid-twentieth century wrote two books of essays on English intellectual history called, respectively, *The Seventeenth-Century Background* and *The Eighteenth-Century Background* as well as two others called, with an understandable aversion to excessive repetition, *Nineteenth-Century Studies* and *More Nineteenth-Century Studies*. Other notable examples of the genre if not the phrase were Tillyard's *The Elizabethan World Picture* and C. S. Lewis's *The Discarded Image*. To be autobiographical again for a moment, as an undergraduate, like many others, I read these works, and learned from them, and enjoyed them, though they were not I think prescribed reading for the Cambridge History Tripos. It was only after the first enthusiasm had worn off that I began to feel piqued. What these authors were doing, after all, seemed to be history. But history, to me and to the faculty which formally instructed me, was not 'background' but 'foreground' – what one would be at. Still, no doubt background and foreground were interchangeable, their back our fore and vice versa, as an historian might illustrate an essay on eighteenth-century social conditions by a reference from Fielding or Richardson. That seemed symmetrical and fair enough, except that in fact there was a crucial asymmetry and it was not fair after all. Because where in the historian's foreground was the literary critic's 'background' brought forward? Only, generally, as the political theorists who it was often implied were not really 'relevant' or 'real history', even though their continuing ability, for mysterious and no doubt historical reasons, to manifest themselves in examination questions meant that undergraduates at least needed to regard them with a certain prudential respect.

The point of this reminiscence is one about the different reasons for attending to texts. The shades of distinction between an historically sensitive literary or philosophical reading, and a philosophically acute or critically sensitive reading of, it maybe the *same* text considered as evidence for a piece of intellectual history, are given by the shifting priority which can be accorded to text or context. The spacial metaphor of background has come to seem inappropriate. Texts both secrete and are woven into the contemporary discourses of which they are part; they are not merely set against them. Nevertheless, it does seem to me that the differences in focus of attention registered by the old spacial metaphor, background and foreground varying according to the nature of one's interest, remain real and legitimate and it is convenient that we should have ways of registering them, without implying that they need be mutually exclusive, even within the same piece of scholarship or analysis. In other words, just as in speaking earlier of history and historians I was claiming a necessary space – not perhaps always readily accorded – for *intellectual* history, so now I have been trying to establish that though intellectual history may sometimes attend to the same texts, and, depending on the kind of text, even up to a point employ the same methods, as other disciplines, it is nevertheless, if it is to be worth distinguishing as a form of enquiry at all, a form of *history*. And one way of describing the kind of history with which it is characteristically concerned is the reconstruction of the intellectual contexts of past texts and utterances.

iv

Whig intellectual history has had, at least since the Enlightenment, a kind of intuitive obviousness. It is natural to human beings to take themselves as their point of reference, and as the organising principle of their history. It is perhaps particularly so when dealing with ideas, which are not bounded in time in the same way as actions or institutions; indeed, we give the latter a kind of afterlife by converting them into ideas. Ideas press their claims on us, urgently in some cases, fancifully in others, for acquiescence or rejection. The *history* of ideas is an afterthought, a product of sophistication and a sense of distance, and, since in some sense we necessarily hold the ideas we do as, at least provisionally, true, the most obvious way of making the history of ideas significant for ourselves is to see it as accumulation, accompanied by possibilities of deviation or heresy: a pilgrim's progress of enlightenment, beset, it may be, by temptation and error.

The antithesis often set against this conception of intellectual history as recording progress or heresy is the German Romantic notion of a past composed of discrete cultures, of a diversity of distinct voices. Yet, at least in England, a conception of successive 'contributions' long remained, I think, the tacit or explicit organising principle of most essays in the history of ideas. What has in more recent years given the practice of intellectual history its distinctive character is a cultivated resistance to this conception of it. Famous examples included the attempt, associated above all with the late Frances Yates, to consider Renaissance thought and early modern scientific ideas in a way which makes their characteristic concern with magic, astrology and alchemy not aberrant, embarrassing deviations from essentially forward-looking intellectual movements, but intrinsic, and essential to a full understanding of their character. Another has been the rewriting, for it is no less, of the history of Anglo-American political thought in the seventeenth and eighteenth centuries, emanating largely, directly or indirectly, from Cambridge, England and Cambridge, Massachusetts. It has involved substantial revision of our views of some of the heroes of the liberal canon. Locke, in John Dunn's *The Political Thought of John Locke*, an account published nearly 40 years ago, came to look distinctly yet plausibly unfamiliar, as a seventeenth-century Calvinist rather than a proto-liberal. J. G. A. Pocock showed in *The Machiavellian Moment* (1975) how different eighteenth-century English political thought began to look when seen in terms not of a prospective 'bourgeois liberalism' but of a late flowering and development of Florentine civic humanism, while Donald Winch in *Adam Smith's Politics* (1978) made use of these insights to revise the hagiography and demonology which concurs in seeing Adam Smith as the prophetic spokesman of nineteenth-century laissez-faire doctrines and to reinstate his thinking in its complex eighteenth-century context. Intellectual context has been the crucial concern here, as has the insistence that meaning is context-dependent, an insistence which was given its most combative and effective statement in a now deservedly famous article by Quentin Skinner in 1969, 'Meaning and Understanding in the History of Ideas'.

But how are we to characterise contexts? Pocock, in his essay 'Languages and their implications' (*Politics, Language and Time*, 1971), spoke most explicitly of this, drawing on and adapting Thomas Kuhn's conception of the role of paradigms in the history of science to reconstructing the diversity of languages, rhetorics or modes of discourse available at any given time, and the manner of their exploitation for purposes of debate and persuasion. Pocock, as I say, speaks of 'languages'. Employed in this way, suggestively and metaphorically, there can be no objection to our doing so. We need ways of identifying relationships in the intellectual life of the past more comprehensive and subtle than the particularist vocabulary which identifies 'utterances', 'arguments', 'concepts', 'texts', 'authors' or 'doctrines'. And there is virtue in the plural – 'languages' – variously available within the same period as vehicles of debate and disagreement. Yet, of course, in devising general terms and metaphors for the coherences we perceive, we run the usual risks of becoming captives to our own concepts and their particular kinds of metaphorical suggestiveness, so that we can respond to no others. When this happens, the concepts we have adopted to help us identify the alienness and distinctiveness of aspects of the past become barriers to our further understanding of it. I have rather more in mind here than just a general and no doubt unhelpful call to intellectual strenuousness and periodic conceptual reappraisal. The general question I am raising is how far what we may call 'the identity of the history of ideas' requires that it should have an agreed and systematically related body of concepts which constitutes the language and identifies the subject matter of the intellectual historian. I want to argue that it does not, and if that is what 'the identity of the history of ideas' demands then I am only a very lukewarm and partial proponent and representative of it.

It seems obvious that we are not, in fact, yet confronted by a fully-worked out theoretical language for a proposed science of cultural or intellectual history, much less an agreed one. What we do have, and for quite a long time have had, is tendencies, dispositions, yearnings even, expressed in different comprehensive labels or metaphors, used not casually and eclectically, but systematically, to suggest how a distinct past set of intellectual relationships must necessarily be seen in order to be understood in its totality. The older English conception of the history of ideas essentially as histories of distinct disciplines with their characteristically Whiggish tendency, depended on a too ready assumption of continuity and ease of access to the past and also a neglect of wider contemporary intellectual contexts; such histories of ideas sacrifice the lateral to the vertical, the synchronic to the diachronic. Here, and with no less wariness, I want to draw attention to the opposite, an assumption that past cultures form coherent totalities, leading to an emphasis on their discreteness, as islands of meaning, distinct both from ourselves and from each other, to the point where it has sometimes seemed difficult to reconcile such cultural models with the notion of history as a continuum, and with their possible interpenetration and transitions. Change becomes characteristically conceived, as Michel Foucault put it, as 'a radical event that is distributed across the entire visible surface of knowledge'.[4] Confronted by such formidable discreteness, the freshman's truism that all ages are ages of transition becomes less a cliché than a subversive thought. Discontinuity becomes the norm, and it, and the problems it poses for our understanding of change, is as apparent in Kuhn's view of the history of science as a succession of self-sustaining paradigms as it is in the older and broader periodisations of *Geistesgeschichte*. It is the difficulty of getting such self-contained cultural totalities on the move, as it were, that gives its opportunity to the *force majeure* of materialist historical explanation; they seem typically to require an external force to make them move. It is clearly not accidental that the native soil of *Geistesgeschichte* and *Kulturgeschichte* is also the fatherland of the materialist interpretation of history. But Marx turning Hegel on his head is surely a cautionary tale for both parties, implying mutual need. The *explanans* of the material forces of production requires among its *explananda* something like the large cultural coherences postulated by *Geistesgeschichte*.

But the general explanation of changes in ideas is not my theme; indeed, I cannot find any reason to suppose that they have a single general explanation at all. The distinction between the history of ideas and other kinds of history rests on practice and the conveniences of the scholarly division of labour, not on ontologically distinct categories of being standing in some consistent casual relation to each other. My concern here is only with how the coherence of ideas as patterns of meaning, and their location in historical periods, have been pictured, and in terms of what models and vocabularies, and with how far the history of ideas may require such models as its objects of attention. The most primitive of these was the one coined in Germany in the Romantic period to express a laudable sensitivity to the cultural diversities exhibited by different epochs in the European past: *Zeitgeist*, the spirit of the age. The inspiration here may have been metaphysical, as in Hegel's systematically worked out conception of history as the successive moments of an active and informing idea, but the model was, or quite soon became, I think, in a loose sense psychological: a cultural epoch was given its distinctive character by the predominance in it of a particular psychological archetype, of which the most famous was 'Renaissance man' with Jacob Burckhardt as his impresario. One understood a cultural epoch as one 'understands' an individual, in terms of consistency of disposition. But there has been a number of other terms advanced, each with a different kind of suggestiveness and derived from different models, not only psychological but metaphysical, sociological or socio-economic, astronomical and, most recently, linguistic. So we are invited to categorise the intellectual life of the past in terms, most loosely, as I have done here, of cultures, but also of ideologies, worldviews or *Weltschauungen*, cultural hegemony, paradigm, episteme, universe of discourse.

The eclecticism of this list seems to me reassuring: not a sign of immaturity in intellectual history but healthy and desirable given the kind of pursuit I take it to be, offering various kinds of suggestiveness but, except for zealots, able to impose none exclusively. But even in their diversity I think we need to be wary of them. The assumption that intellectual history needs a special and systematic theoretical vocabulary for the entities and relationships with which it deals is implied by two closely related beliefs. The first is that the portions of the intellectual life of the past attended to are to be thought of as alien, closed, hard-edged. The second is that they were also highly coherent, in ways unperceivable by those implicated in them, so that we can understand their intellectual lives, in them, so that we can understand their intellectual lives, in their totality, and perhaps in connection with other aspects of their social organisation, as they themselves could not. For the hard-edged, closed totality that we claim to perceive we need, clearly, a model. But that is surely the wrong way of putting it. Our belief that we perceive such a totality is most often a consequence of the possession of the model, and it is typically one we bring ready-made to some aspect of the past, rather than one we construct, *ad hoc*, in our negotiations with it. There is, of course, a powerful and rather fashionable inclination to say that the former is inevitable; to stress the shaping power of the concepts and presuppositions that we necessarily bring to the past, to the point where it seems that in what we take to our encounters with it we hear only the echoes of our own voices. I do not believe that this is inevitable, though I think something like it can and does occur. To put the point merely as an affirmation, I believe that our relation to the discourse of the past can be a kind of negotiation, in the sense that it can be two-sided; that we can listen as well as speak, and that even in our necessarily selective and partly ignorant or inattentive or impatient listening we can sometimes hear what we did not expect. And it is in recognising the ability of the past to surprise us, in attending above all to what does not fit our preconceptions, that we can learn. We cannot be without preconceptions, but they can be more or less insulated from surprise, and the more systematically technical we try to make them the more effectively insulated they are likely to be, and the deafer we shall be to anything we did not already intend to hear.

I must guard against misunderstanding here. I am not saying, absurdly, that in our renderings of past discourse we should be restricted to the vocabulary of that discourse, though I would always feel more comfortable in remaining closely in contact with it. Intellectual history is not parody. Nor am I denying our right to coin, or borrow from elsewhere, concepts for interpreting it, making connections and analogies that might otherwise have escaped us, and helping us better to understand the tacit rules and conventions and limitations of which speakers in the past were not, or were not habitually, conscious, and of which they did not therefore explicitly speak, as we are not usually consciously aware of, nor do we usually feel constrained by, the grammar of our own language. Not least, I do not want to deny that there can be important benefits in learning sometimes to see the apparently familiar as strange, including perhaps the discovery that it is less familiar than we had thought. I do not want to deny any of that. What I want to claim is something less, but still important, namely, to reiterate the point, that there not only is not – which seems at the moment to be simply true – but also that it is better that there should not be, any single, unified methodology, conceptual scheme, theoretical language whose use definitively characterises intellectual history and its practitioners.

v

I may quite properly be asked what general description best fits what I think I myself do, and how, in general, we may extend and deepen our familiarity with the intellectual life of the past and so enlarge the boundaries of the intellectual parochialism of the present. In reply I would offer two analogies. The first is that of eavesdropping on a conversation: not wholly opaque or impenetrable or alien, though conducted, of course, without reference to ourselves, but also tantalising in its obscurities, its allusiveness, its elusive implicit assumptions, its constant hinting at larger unseen contexts. To practise intellectual history, as it seems to me, requires not knowledge of a model or recipe, but simply patience, alertness and persistence in our attention to the conversations of the past, trying to understand them as we learn a natural language, circularly, coming to understand the components by their contexts and building up our sense of context as we identify its components. Of course there is a large and impressive body of philosophical and linguistic theory about what we do when we do such things, when we understand and transmit meanings, but they are not my concern. The theories, I take it, are accounts or explanations of what we do, they are not recipes for doing them.

The other analogy I would offer is that of translation. One of its virtues, I think, is to draw attention to a specific inadequacy in another account which has sometimes been offered of intellectual history, as the 're-creating' of or 're-experiencing' of past thoughts. This is a rather unclear notion in any case, but the particular inadequacy to which the analogy of translation draws attention is the neglect of the question of transmission. To think past thoughts, if that is the right way of putting it, is only half the process; there is little point, even supposing it to be possible, in as it were, retrospectively adding to the population of a past society by one. The point is to be able to talk about it, to be a medium between two worlds. Translation reminds us of this because it is never just translation from but also translation *into,* and this is a curiously neglected aspect of historical writing. To and for whom, when we try to do intellectual history, do we translate? The answer is to at least a limited extent, a choice, not just a given. Presumably to oneself and one's readers or pupils: more generally ourselves – who else? But who are 'ourselves'? The twenty-first century, even in Britain alone, is a fairly culturally bewildering, heterogeneous and sometimes rather sectarian place. In choosing the idiom and vocabulary into which to try to render one's understanding of past utterances, one in a sense chooses both one's past and one's present, declaring what one finds of interest in the former and also whom one wants to address in the latter and now. And if anyone wants, in the extended sense of the word which is, or at least used to be, fashionable in making this kind of point, to call the choice necessarily a political one, I shall not disagree. If, as in this kind of rhetoric, everything is political, there is no point in arguing that something in particular is not.

This is not the place to consider specific advantages and disadvantages of particular kinds of idiom into which one might choose to translate. I have only time to recommend the virtues of one, the most obvious. In doing so I am, of course, aware of all the inhibitions which have grown on us about tacitly assuming the authority of neutrality of 'ordinary language' or in making claims, in Matthew Arnold's vein, to some kind of cultural 'centrality'. Our confidence in the possession of a common culture is not what it was. But we do, it seems to me, still have a usable notion of the common, educated lay speech, the 'ordinary language' of our time and country. And it also seems to me that this, as far as possible free of technicalities and neologisms, is likely to prove, overall, not only a more widely usable but also a more sensitive and flexible receiver of the overheard conversations of the past, better because richer, more heterogeneous, and hence more variously attentive, than the tight, imperious, systematically imposed categories and definitions insisted on by a consistently selective, self-conscious methodological doctrine or mode.

What I have been talking about is two different and opposite ways of limiting and impoverishing our negotiations with the reflections, arguments, questionings conducted by people in the past. The first kind of avoidable limitation is the way of anachronism, of the too ready assimilation of these things into our categories, or their too facile and unselfconscious conscription into some academic discipline or 'great tradition' to which we choose to attach ourselves. The other kind of impoverishment, surprisingly perhaps, is the opposite: an over-schematic and hard-edged packaging of our perceptions of the alienness, the otherness or historicity of the past. These two opposite poles represent I think two characteristic ways of refusing to enter into genuine negotiation with what is not ourselves, refusing to hazard any fragment of our own complacency as part of the negotiation. The first corresponds, in contemporary life, to the tendency to deny cultural difference, to assimilate others to versions of ourselves, and to assume that if they seem not to yield to such assimilation it must be because they are inferior, backward, childish, retarded. The opposite and now more fashionable kind of refusal exaggerates difference or alienness, treats it as fixed, definitive, exhaustive, and labels it like a museum specimen seen under glass, reducing human beings to whom we purport to refer to a cultural category. But any person or culture or society or period is more various, elusive, transitional than any other stereotype we can devise for him or it. What we have to accept in the history of ideas, as in other branches of scholarship, is the endless alertness required in our attempts both to accept the complexity of the past and to express our sense of its coherences, to acknowledge its distinctiveness without losing sight of its human energy and variety, so that we do not reduce its inhabitants to implausible puppets, manipulated without reminder by whatever model we devise for understanding but also for distancing them, to see them simultaneously as potentially useful guides and as tiresome officious couriers, trying to package and standardise our perception of the alien. There is no definitive conceptual resting-place, no ultimate methodological vantage-point. And this, it seems to me, should neither surprise nor dismay us, because we do not have these in our relations with and attempts to understand the human beings with whom we share the world as contemporaries. Why should we expect to enjoy our relations with the dead more complacently?

notes

1. A. E. Houseman, *Selected Prose*, ed. J. Carter (Cambridge, 1962), p. 12.
2. F. W. Maitland, *The Life and Letters of Leslie Stephen* (London, 1906), p. 283.
3. E. P. Thompson, *The Making of the English Working Class* (Harmondsworth, 1968), p. 13; S. Collini, 'A Nebulous Province: the science of politics in the early twentieth century' in S. Collini, D. Winch and J. Burrow, *That Noble Science of Politics. A study in nineteenth-century intellectual history* (Cambridge, 1983), p. 377.
4. Michel Foucault, *The Order of Things. An archeology of the human sciences* (New York, 1972 [orig. 1969]), p. 217.

2

intellectual history in britain

brian young[1]

Intellectual history has often seemed rather a foreign activity in British historical circles; some of its major practitioners in the twentieth century were themselves refugees from cultures in which the history of ideas was altogether more widespread a scholarly activity than it has ever proved to be in Britain. The first half of this chapter will explore the immediate post-Second World War renaissance in the practice of intellectual history in Britain, and the second half will examine how it has changed as an activity since the days when such grandees as Isaiah Berlin and Arnaldo Momigliano dominated the subject. There is a useful symbolic break in this chronology, originally appearing as recently as 1997, when the views of an unusually dominant refugee scholar, the late Sir Geoffrey Elton, who was a strikingly assiduous opponent of intellectual history from the perspective of an ultra-empirical political historian, were subjected to a searching refutation by Quentin Skinner, the most recent of his successors as Regius Professor of Modern History at Cambridge.[2] How was it that intellectual history moved from being the subject of professorial denunciation in Cambridge to a primary subject of later professorial practice there? What does this correction of prejudice reveal about the wider standing of intellectual history in British academic and cultural life as a whole?

i

Writing within half a dozen years of each other, two refugee scholars (one of whom had left revolutionary Russia with his parents when a child, the other of whom had been rescued from Mussolini's Italy), made contrasting statements about the position of intellectual history in England, their adopted country. Isaiah Berlin declared in 1966 that:

> Intellectual history is a field in which English writers, in general, have taken less interest than those of other countries. There are notable exceptions to this rule; but they are few. The history of English thought, even in the nineteenth century, when it had greater influence than that of any other country, still remains to be written.[3]

For Berlin, intellectual history had attained greater authority on the Continent than it had ever had in England; he worried more over its meaning when writing about Russian intellectual history than he ever did when considering English matters, about which he wrote remarkably little.[4] Writing in 1972, Arnaldo Momigliano, the Piedmontese historian of antiquity and humanist polymath, observed that it was only in the 1930s that the English – probably at the behest of Lewis Namier, an earlier intellectual émigré – had come to believe that 'the history of ideas was an unBritish activity', observing, by contrast, that during his student years in Turin during the 1920s,

> the history of ideas was the speciality for which English historians were most famous. This reputation went back to the days of Grote and Lecky, Freeman, Bryce and Flint. There were few books to compete with Leslie Stephen's *History of English Thought in the 18th Century* or with J. B. Bury's *A History of Freedom of Thought* and *The Idea of Progress*. Lord Acton managed to become famous for a book on liberty he did not write.

Momigliano brought the catalogue up to date with reference to works produced by English scholars during his own student days. Things had begun to change, even then, however, and he noted the high and competing esteem in which German *Ideengeschichte* was held in Turin's faculty of arts in the 1920s and 1930s.[5]

Why did these two historians of ideas diverge so fundamentally on this issue? In part because Berlin, as he constantly reminded his readers, was never an historian by training or even by inclination; he was also indebted to the anti-intellectual historian Lewis Namier in his early and formative understanding of history as the subject was practised by contemporary academic historians in Britain when he himself was beginning to practise his first academic vocation as a philosopher. He was, however, properly critical of Namier's lack of interest in ideas, and situated this tendency in his thinking with elegant accuracy as a product of the Vienna that was at the centre of 'the new anti-metaphysical and anti-impressionist positivism', the Vienna of the Logical Positivism that so influenced Berlin's philosophical colleague, A. J. Ayer.[6] Berlin had subsequently become interested in the German Renaissance of the late eighteenth and early nineteenth centuries. He discerned in such thinkers as

Herder early promoters of an ancestral variant of his own commitment to pluralism in intellectual, cultural and moral life, albeit one which ended, in the case of other thinkers of the German Renaissance, in the nationalism to which Berlin remained resolutely opposed, and which frequently invoked the spectre of race, a category about whose mishandling Berlin was never less than humanely scornful.[7] For Berlin, reflection on the thought of what he called the Counter-Enlightenment usefully complicated the otherwise predictable contours of European liberalism, a political position to which he was personally committed.[8] If temperamentally closer to the Enlightenment, Berlin was able to use one of the creations of the Counter-Enlightenment temperament, what Herder called *Einfühlung*, the capacity to empathise, to feel one's way into the thoughts and feelings of others, particularly those from whom one is naturally distant culturally.[9] His perspective, despite his veneration for John Stuart Mill, was always continental, and occasionally Russian, in tone; England, the land of practical liberty, was not a land of ideas as such for Berlin.[10] This suited him personally, but it did lead to a mild distortion in his depiction of the history of ideas in which the contribution of nineteenth-century English historians and thinkers, of the kind praised by Momigliano, was marginalised by Berlin in favour of continental thinkers and activists.

As Berlin himself stated the matter, he attributed his interest in ideas to being Russian in origin, witnessing through his examination of the Russian experience of the twentieth century a society disastrously given over to the pursuit of revolutionary ideas; his own values, on the other hand, he saw as being quintessentially English, the empirical antithesis of an addiction to abstraction. He characterised this balance with his usual economy: 'An effective antidote to passionate intensity, so creative in the arts, so fatal in life, derives from the British empirical tradition.' In common with Momigliano, he was also aware of how his being Jewish complicated his sense of being a Russian by origin and a Briton by adoption, inclination and commitment.[11] Questions of Jewish identity preoccupied Berlin, as can be appreciated in his magnificent essay on the negotiations with their Jewish roots undertaken by those unlikely contemporaries, Karl Marx and Benjamin Disraeli (it was typical of Berlin to link such disparate thinkers together so productively and imaginatively; he liked complementary contrasts, as in his famous essay on 'The Hedgehog and the Fox', concerning thinkers motivated by one big, all-encompassing idea, in contrast with those who used a variety of insights in their cumulative understandings).[12]

Momigliano, an historian of scholarship, was much more aware of the value of the work undertaken by the likes of the British historians whose works he had cited in his 1972 essay. He was also more critical of German scholarship, for personal reasons – his parents had died in concentration camps – if no less admiring of it than was Berlin.[13] He liked to emphasise the centrality to the history of classical scholarship of such English scholars as Gibbon and

George Grote, his nineteenth-century predecessor as professor of ancient history at University College, London, and, in the process, extended the reach of Anglophone culture into genuinely European scholarship.[14] Momigliano and Berlin deeply admired one another's work, and Momigliano was strikingly perceptive in noting the appeal of Vico and Herder to Berlin in a way that he found personally sympathetic:

> He must have found in Vico and Herder a welcome alternative to that analytical philosophy – carefully eschewing any problem about poetry and myth, indeed about history in general – with which he grew up in Oxford. More specifically he must have found in Vico and Herder confirmation and support in his own lifelong fight for cultural pluralism and respect for minorities (including his own – should I say our own – the Jewish minority).[15]

Berlin found this review of his work suggestive enough to reply to one of its central contentions (that his eighteenth-century subjects were relativists) in a typically astute essay.[16] Before opening up this contrast in perceptions of English intellectual history between Berlin and Momigliano, it is well to emphasise the fact that it was one created by émigré scholars, whose contributions to British intellectual history have been immense. Not that all students of history and the social sciences have been so impressed by this contribution, and even so distinguished a scholar as Berlin has been slighted by critics who have not shared in his detestation of state planning and 'positive' freedom.[17] The Marxist critic Perry Anderson notoriously claimed in a polemical essay published in 1968 (the most recent year of European revolt), that such émigrés had assimilated too readily to the limitations of the local culture, and that they brought to their new home none of the intellectual vivacity that such Marxist *savants* as Theodor Adorno and Max Horkheimer had taken with them from Europe to the United States. Instead, they accepted knighthoods and celebrated English virtues; Berlin was dismissed, alongside Karl Popper, as being a fluent ideologue, whilst Ernst Gombrich, whose work in art history was strongly indebted to intellectual history, was witheringly praised as an 'honourable but limited' pioneer in his field.[18] Berlin would have had much to say to a revolutionary enthusiast of the failed revolution of 1968, just as he and Namier had had much to say about the failed European revolutions of 1848.[19]

It was the calamities of the 1930s, however, and not the differently troubled atmosphere of 1968, that shaped the outlooks of scholars such as Berlin, a period when German patterns of scholarship were, understandably, at their least attractive to non-German scholars. It was to this atmosphere that Momigliano pointed when he alluded to an important strand of Cambridge intellectual history in drawing his distinction between *Ideengeschichte* and English traditions in the history of ideas. On the one hand, he pointed to the work of the Germanist Eliza M. Butler, a fellow of Newnham College, especially as represented in what he called 'her singular criticism of German humanism', *The Tyranny of Greece over Germany* (1935). On the other hand, was the work and example of Sir Herbert Butterfield, whose attitude to intellectual history was decidedly complex, as will become apparent later in this chapter.[20] Butler's study of humanism in modern Germany demonstrates how rooted in the 1930s were the conceptions of national tendencies in intellectual history that Berlin would deploy in his writings well into his final years in the 1990s. Aware as she was of the rise of National Socialism, there is something of the catastrophic in Butler's account, as is apparent from the opening pages of her study:

> For the Germans cherish a hopeless passion for the absolute, under whatever name and in whatever guise they imagine it. The Russians have had stranger visions; the French have shown themselves more capable of embodying abstract ideas in political institutions; but the Germans are unique perhaps in the ardour with which they pursue ideas and attempt to transform them into realities. Their great achievements, their catastrophic failures, their tragic political history are all impregnated with this dangerous idealism. If most of us are the victims of circumstances, it may truly be said of the Germans as a whole that they are at the mercy of ideas.[21]

Sandra Peacock has discussed Butler's negative preoccupations with German culture;[22] was it this negativity that made Butler's work so attractive to Momigliano? What is certain is that her level of deeply informed cultural generalisation was akin to that indulged by Berlin in his work. In this sense, Butler belonged to a generation that was preoccupied by the political distortion of ideas, and scholars at Cambridge were at one on these matters with Berlin's Oxford generation.[23]

It was largely on the political left in Britain in the 1930s that the study of the history of ideas made its greatest impact. Goronwy Rees, a colleague of Berlin at All Souls College, Oxford, and who later played a part in revelations about the Cambridge spies Kim Philby, Guy Burgess and Donald Maclean, had started research on Ferdinand Lassalle, Marx's rival, at the time when Berlin was researching his own short study of Marx.[24] Rees's memoir of his time in the city of Berlin in 1934, when he had initiated his quickly abandoned research (he had been discouraged by the Warden of All Souls from starting alternative work on 'an intellectual history of young Marx'), speaks volumes about the political commitments which informed so much of the intellectual history written (or at least conceived) during that tempestuous decade.[25] No matter how provincial English intellectual history might have ultimately appeared to Berlin, it was no less politically engaged than anything written on the Continent, even if the registers in which it was conducted were admittedly infinitely less combustible. Likewise, in the America where Berlin had worked at the British Embassy during the Second World War, Edmund Wilson would produce a history of socialist thought which contains acute analyses of many of the thinkers, from Vico to Marx and Bakunin, about whom Berlin would come to write. Wilson's study of the history of socialism in Europe and Russia, *To the Finland Station* (1940), was written from a political position well to Berlin's left, a position many intellectually engaged British writers also occupied in the 1930s and 1940s.[26]

The work of one such scholar, Anthony Blunt, the last of the Cambridge spies to be exposed (as recently as 1979), repays attention in this respect; it also demonstrates how intellectual history so frequently overlaps with the history of art. It should also be noted that the exile of an institution devoted to art and cultural history, the Warburg Institute, from Hamburg to London in 1933 has had an enormous impact in the field of intellectual history, notably in the work of Blunt's London University colleagues D. P. Walker and Frances Yates.[27] This overlap is most evident in Blunt's much later study of Nicolas Poussin, the product of a lifetime's labours, in which he read the artist's work as embodying the neo-Stoicism of the seventeenth century: a concern with ideas was seen to be informing the secret life of an artist and his circle, a motif which had likewise been central to Blunt's own career as a former Soviet agent turned-Surveyor of the Queen's Pictures.[28] An historian who pondered the atmosphere of the Cambridge in which Blunt and his associates were educated was Noel Annan, who had served as Provost of King's College, Cambridge before moving as Provost to University College, London.[29]

Annan was deeply affected by his experience of the Second World War, in which his work in intelligence took a rather different form from that of Blunt, Burgess, Philby and Maclean. Annan not only took part in war-time intelligence gathering, but he also played a prominent role in the post-war re-establishment of democratic politics in what was to become West Germany, a process about which he wrote a typically elegant, characteristically mandarin account.[30] Already fluent in German before the outbreak of war (it was this ability that led to his original recruitment), Annan became deeply engaged by the German Renaissance of the early nineteenth century which also preoccupied Isaiah Berlin. This informed Annan's seminal account of the intellectual contexts in which the nineteenth-century man of letters Leslie Stephen (the pioneer of intellectual history in Britain) had made his career, a study first published in 1951 and revised in 1986, a book which counters, to some degree, Berlin's claim that no intellectual history of nineteenth-century Britain written by an Englishman had appeared by the mid-1960s.[31] Berlin was one of Annan's intellectual heroes, and in his survey of the intellectual world in which Annan had grown to maturity, he concluded that Berlin had written 'the truest and most moving of all interpretations of life that my own generation made'.[32] Elsewhere, he described Berlin as a 'magus', celebrating his use of intellectual history as a means of examining and complicating the ideas by which we live our lives in the present.[33]

Berlin was indeed extremely aware of the insidious way in which ideas formulated in the nineteenth century had been perverted in the twentieth century into ideologies which excused the torture and murder of millions, from Stalin's Gulags to Nazi concentration camps.[34] His essays reinforce this connection again and again, nowhere perhaps more clearly or more adroitly than in an essay on the counter-revolutionary thinker Joseph de Maistre, largely written in 1960, just 15 years after the end of the war. It is an astonishing essay, not least in the manner in which it presents de Maistre with imagination and understanding; Berlin was seemingly able to enter the minds of even the most personally unsympathetic of figures in order to do full justice to thinkers who were not always willing or able to do full justice to those they themselves had all too easily condemned.[35] Berlin's liberalism affected the way in which he approached his subjects, as well as in his choice of subjects; clearly, he infinitely preferred the world of the Zionist Socialist Moses Hess to the reactionary world whose sanguinary injunctions were promulgated in the writings of de Maistre, but both needed to be understood if anything of the undoubted worth of the social vision of Hess (whom he considered a greater political prophet than Marx) was to be realised.[36] Berlin was concerned to understand the enemies of freedom in order to defend it as strongly as he could in the Cold War environment of the 1950s through to the 1980s, and he had used the opportunity of radio talks in 1952 to alert listeners to the siren songs of such opponents of liberty.[37] Similarly, his Mellon lectures, given in America in 1965 on the genesis, development and legacy of Romanticism, pointed to the political and philosophical ambivalence of that peculiarly charged moment in European history.[38]

A figure about whom Berlin did not write, but who incarnated a reaction within the German tradition against the dangers of populism and of political centralisation and demagoguery, was Jacob Burckhardt, about whom Momigliano wrote briefly but tenderly.[39] Burckhardt was also a hero to another historian who worked in intelligence during the Second World War, a period in which he made himself fully conversant with German intellectual history. Hugh Trevor-Roper, a student of Christ Church who was later to become Regius Professor of Modern History at Oxford, wrote on an astonishing array of matters, and he came increasingly to favour intellectual history in the long high period of his intellectual maturity during which reflection on Burckhardt became ever more important to him.[40] Trevor-Roper had a natural flair for historiography, absorbing stylistic lessons that made his prose a worthy idiom for his masterly short studies of Gibbon and Macaulay and other historical prose masters.[41] Even his study of *The Last Days of Hitler* used a Tacitean style when commenting on the pervasive decadence of the closing moments of the Third Reich. His first book, a lengthy study of Archbishop Laud, also contained an element of intellectual history, as he subjected his subject's religious ideas to amusingly anti-clerical analysis.[42] Laudianism would recur as a subject in his essays in intellectual and cultural history.

Trevor-Roper's ambition for the study of ideas in their historical context which he always considered alongside the high culture in which they emerged is made plain in the preface to his collection from 1985, *Renaissance Essays*: 'my principal interest has been in intellectual and cultural history which I have tried to see not in isolation but in its relation to, and expression in, society and politics: in the realization of ideas, the patronage of the arts, the interpretation of history, the social challenge of science, the social application of religion.'[43] In an earlier collection, *Religion, the Reformation and Social Change* (1967), this deep connection between intellectual and other forms of history had been made evident, and it continued to be so in later collections of essays, mostly concerned with sixteenth- and seventeenth-century subjects; a late essay on Gibbon revelled in the comedy of the great historian's relations with John Pinkerton, the type of scholarly near charlatan who so appealed to Trevor-Roper's sense of the comic in intellectual history.[44] It was a sense of comedy that was often told alongside the bitterness of the past; one can read Trevor-Roper's comments on the horrors of the Thirty Years' War as paralleling those experienced in the same theatres of war (especially if one includes the English Civil War as part of the Thirty Years' War) during the Second World War. Experience of the Second World War had led to Trevor-Roper insisting on the inseparability of English from European history, and also on the longevity of ideas, particularly pernicious ones. This was stated with lapidary economy in the introduction to his *Catholics, Anglicans, and Puritans* in 1987 at exactly the moment when Berlin's essays were being collected to demonstrate similar contentions:

> But how can we speak of 'English intellectual history'? Intellectual history can never be pursued in isolation. It is conditioned by its social and political context, quickened and distorted by events. Nor can it be localized. Though local conditions may give them a particular direction, ideas cannot be confined in space. They overflow national boundaries. They also overflow the boundaries of time. Ancient philosophies, rediscovered, are found to possess a disturbing vitality even in modern times.[45]

The demands such study makes of the scholar, and the benefits accruing from such dedication to those demands, will become apparent once again when his work on the seventeenth-century physician Sir Theodore de Mayerne is published in 2006.

Trevor-Roper practised a rather different version of intellectual history from that propounded by Isaiah Berlin. Indeed, Berlin's work can be seen as the very type of the history of ideas approach to such issues: his was philosopher's history, a concern with individual thinkers and the multiplicity of ideas they discussed, a sort of continuous dialogue over the ages. Reading philosophy in Oxford entailed study of the history of ideas; this was both philosophers' history, in the sense that philosophers studied it for their own purposes, and philosophers' history in the sense that it was very much about what philosophers had thought, and how this related to what other philosophers had thought. This was likewise largely true of the study of the history of political thought as practised by both Berlin and his successor in the Chichele Chair of Social and Political Thought at Oxford, John Plamenatz, a Montenegrin, and also, as demonstrated in his studies of Marxism, by their All Souls colleague, Leszek Kolakowski, a disillusioned refugee from Polish state communism.[46] Writing of Vico, Hamann and Herder, Berlin had insisted that 'the importance of past philosophers in the end resides in the fact that the issues which they raised are live issues still (or again), and, as in this case, have not perished with the past societies of Naples or Königsberg or Weimar, in which they were conceived.'[47] In contrast with this approach, Trevor-Roper practised intellectual history, placing ideas in their fullest possible context, detailing how they affected and reflected the broader concerns of the societies in which they came about and where they either flourished or perished. It is interesting to reflect that the essay by Berlin which contains the most conventionally historical scholarly apparatus, 'The Originality of Machiavelli', and which places its subject most carefully in context, contains an acknowledgement to Trevor-Roper, who drew to Berlin's attention the supremely historical insight that Machiavelli's admired examples were wholly or in part mythical figures.[48] Trevor-Roper ploughed a somewhat lonely furrow in Oxford, in that little intellectual history was undertaken there when he was professor; certainly, it was far less significant in historical studies there (with the exception of the work of such medievalists as Sir Richard Southern and such students of late antiquity as Peter Brown) than it was in Cambridge, where Trevor-Roper finished his academic career as the Master of Peterhouse, whose Laudian associations he relished detailing in essays written during his mastership.[49] However, just as Quentin Skinner took up his tenure of the Regius Chair in Modern History at Cambridge, so the Cambridge-educated R. J. W. Evans succeeded Sir John Elliott in Trevor-Roper's old chair at Oxford, and Evans's own expertise in the religious and cultural history of early modern Europe marked the return of intellectual history to a position of prominence within the university. Evans's seminal study of the cultural and intellectual world of Rudolf II, published in 1973, is one of the major works in intellectual history produced in Britain since the war, and his own interest in language and linguistics, evinced in his inaugural lecture as Regius Professor, has paralleled and complemented Skinner's work in this field.[50]

ii

The deep roots of intellectual history at Cambridge have been emphasised by Mark Goldie in a searching essay concerning J. N. Figgis, an Anglican priest and the premier British historian of political thought in the 1900s, and himself an editor, with F. W. Maitland, of the papers of Lord Acton. Figgis's deep studies of the relationship between Christianity and early modern political thought laid strong foundations on which later Cambridge-educated historians of political thought, from John Burrow to Quentin Skinner, have built.[51] Maitland himself was another leading historian of political and legal thought whose legacy has recently been considered by David Runciman and Magnus Ryan, two younger products of the 'Cambridge School' of intellectual history.[52] Although themselves close to Acton, neither Figgis nor Maitland acquired disciples, and something of the driven individual is apparent in their writings, as it is in the work of a Cambridge intellectual historian who came of age alongside a remarkable generation of historians in the 1940s. Duncan Forbes, who had won the Military Cross during the Italian campaign in the Second World War, pioneered the pursuit of intellectual history at Cambridge; his special subject on the Scottish Enlightenment was the first intellectual history option to be adopted by the Cambridge History Faculty. John Robertson has pointed to the intellectually influential undergraduates who studied the subject with Forbes: John Dunn and Quentin Skinner, who remained in Cambridge, and José Harris, who moved to Oxford after a period of teaching at the London School of Economics.[53] Forbes was too personally idiosyncratic to found a school, but he signally helped to shape the young minds that would go on to create the contextual school of Cambridge intellectual history.

Forbes's work repays the close attention it demands. His vigorous study of *The Liberal Anglican Idea of History* is a landmark study of the English reception of the German Renaissance, and of the part played in that development by the ideas of Vico; its compressed, demanding style contrasts with Berlin's clear, whirlwind declamations, but both authors offer perspectives on the thought of the nineteenth century that can be combined to great interpretative effect. In a series of studies in the *Cambridge Journal*, Forbes excavated the ideas of Sir Walter Scott and James Mill, and he initially explored the 'Scientific' Whiggism of the Scottish Enlightenment in a manner that would later bear fruit in one of the greatest works of intellectual history to be produced by a British historian in the twentieth century, *Hume's Philosophical Politics* (1975).[54] In a posthumously published essay, Forbes emphasised the isolation attendant on the rigorous application of the historical intelligence to intellectual history, an essay that carries both great conviction and great force.[55]

Forbes was the senior figure in an outstanding generation of Cambridge historians of political thought that includes the late Maurice Cowling and J. G. A. Pocock, as well as Charles Parkin, the author of a valuable study of Burke, and Eric Stokes, now rather better known as an historian of the British Empire and Commonwealth, but also the author of *The English Utilitarians and India*, a significant study which brilliantly united intellectual with imperial history.[56] The pre-eminent figure in that generation is Pocock, but attention has also to be paid to the idiosyncratic figure of Cowling if sense is to be made of the later products of the Butterfieldian strain in English intellectual history referred to by Momigliano. Butterfield himself wrote mostly on historiography, and it is in this field, along with a volume on the history of science, that he is now best known to intellectual historians. He is also widely known as the author and opponent of *The Whig Interpretation of History* (1931), a critique of present-centred historians who viewed the history of Britain as a chain of Protestant-cum-secular progress which led, inexorably, to the triumphant celebration of the here-and-now, a chain of reasoning not unfamiliar to us now who live in a political climate in which much is made of 'modernisation' and of the claims of the present (and the future) over supposedly complacent cultures of conservatism.

It was as an apostle of a particularly mordant brand of conservatism that Cowling chose to position himself in English intellectual life. He wrote a study of John Stuart Mill in the early 1960s that provoked much resentment as he argued that Mill sought to secularise English intellectual life in order that his own brand of liberalism should prevail over religious understandings of both public and private life.[57] He then devoted himself to chronicling the manoeuvrings of English political elites from the 1860s to the 1940s, an activity in which the assessment of political success overrode considerations of political thought and doctrine, but it was to the doctrines that were at play during the *longue durée* of the political atmospheres so described that he finally turned in the three volumes of his own political and religious statement, *Religion and Public Doctrine in Modern Britain* (1980–2001). These are not easy volumes, either to read or to comment on, but they richly repay close attention, not least as they offer a perspective on intellectual history which is totally unlike that promoted by the likes of Berlin and Annan, intellectual aristocrats against whose dominance in public discussion Cowling liked to thrash out from time to time, just as Butterfield had attacked the complacencies of the liberal historians whose Whig views he denounced from the vantage point of a religiously-informed and decidedly technical understanding of history as both a process and an activity. If one is to read intellectual history as contested territory, as one must, then acquaintance with the writings of Cowling is strongly to be recommended. His was a unique stance in modern intellectual life.[58]

Such is the peculiar style and stance of Cowling that any attempt to do justice to his *magnum opus* is bound to fail, not least as so much of it is taken up with intensely concentrated study of the life and writings of a huge array of British politicians, historians, theologians, philosophers, and men of letters that one would be offering a condensation of a condensation when seeking to offer an insight into the intellectual tendencies of his three volumes. One route to the organising mentality behind the structure of the books is to consider just how concentrated on Cambridge is so much of his positive doctrine, as can be appreciated by two negative appreciations in a chapter of the first volume concerning the conservative thinking of many fellows of Peterhouse, where Cowling was a fellow from the mid-1960s. First is his distillation of the character and political instincts of Sir Denis Brogan, Professor of Political Science at Cambridge between 1939 and 1968:

> Brogan was sensitive, tough-minded, emotional and deeply sympathetic. He drank a great deal, and gave drinking an ideological significance. He had a cant of his own – the cant which tough-mindedness becomes when it is merely a posture. But in general this was overlaid by a vigorous outpouring of contemptuous wit which disposed of Virtue in any ordinary liberal sense.[59]

Similar, if rather less positive treatment is accorded to the liberal Conservative historian, George Kitson Clark, a fellow of Trinity College: 'The trouble with Kitson Clark, if it can be put thus crudely, was that he was untouched by relativism. Instead of irony and complication, he left an impression of simplicity and solid worth.'[60] Nonetheless, at its best, Cowling's study fused institutional with intellectual history in an exemplary manner, and he could convey an atmosphere with great economy, if often with relentless repetition, as in his evocation of the tone of the politician Enoch Powell's pre-war poetry: 'It expressed the position of youth and had an eschatological overtone characteristic of Housman's repressed tombstone emotion. It registered the resigned, masculine gloom of the Trinity ethos into which he had been inducted.'[61]

Cowling's *Religion and Public Doctrine in Modern England* is, fundamentally, a study in intellectual history, an examination of the consequences for public thinking in England of the decline in religious belief from the close of the eighteenth century onwards. His instincts and inclinations, however, remained very defiantly those of a political historian, and it is typical of him to say, in relation to the writings of the literary critic Terry Eagleton and the philosopher Roger Scruton, that he declined to identify himself with either because he believed that 'both take thought far too seriously'.[62] He identified with the two anti-intellectual political historians, Elton and Namier, of whom Quentin Skinner was markedly critical. There is subdued praise in Cowling's reference to Elton's 'exuberant ... Tory positivism' and in his claim that Namier's complicated variant of Zionism offered 'a singular example of the rough, paradoxical conclusions which can be drawn from modern, post-liberal illusionlessness.'[63] It ought to come as no surprise, then, that Cowling was critical of Skinner, chiefly as a result of Skinner's perceived lack of interest in religion as an historical phenomenon, and also because of his publicly stated distrust of Conservatism; his remarks on Skinner are dyspeptic to a degree, evidence of how distant Cowling felt from the Cambridge School's approach to the pursuit of intellectual history since the days of his own intellectual coming of age during the heyday in Cambridge of Butterfield's astringent version of Christian scepticism.[64]

Cowling's criticisms of Skinner are of a piece with the all out assault on the thought and writings of Isaiah Berlin which follow on from his wilfully complicated assessment of Namier. Berlin, whose mind is described as being 'smoother' than Namier's, but 'very much less powerful', was tacitly rebuked for allegedly having led the life of a celebrity when spending time in Kennedy's America, for misunderstanding England, especially English 'moral conservatism and low-keyed mistrust of the higher thought', and for having had 'a defective sense of the ubiquity of force, reaction, accident and institutions.'[65] In short, Berlin was dismissed for not being the sort of political historian that his friend Namier had been, and his interests in ideas as political tools were consequently supposed to have precluded him from properly understanding the political life of the societies he wrote about with what Cowling considered to be an ill-placed confidence. What Anderson and the polemicist Christopher Hitchens did to Berlin's reputation from the position of the New Left, Cowling sought to do from his own complicatedly right-wing perspective.[66] In order properly to appreciate the strengths of Berlin's writings, it is necessary to acquaint oneself with the full range of the thinking of his frequently vociferous opponents. In so doing, one becomes aware of a particularly highly-charged phase in modern British intellectual history that has yet to be chronicled.[67]

One of the many Cambridge figures considered by Cowling was Walter Ullmann, about whom he declared (in a manner that reinforces the centrality of émigrés to the argument of this chapter) that 'Ullmann is a fine example of a recurrent English phenomenon – the European scholar who takes the English out of themselves.'[68] As is demonstrated in this volume in the chapter by Mishtooni Bose, the work of medievalists is very much *sui generis*, and is frequently much more interdisciplinary in character than is that practised by other historians, and it is also often concerned with many of the interests that students of later periods would characterise as intellectual history. The primary actor in this role at Cambridge was Ullmann, a refugee from post-*Anschluss* Austria. Trained as a legal theorist, Ullmann transformed the study of medieval political theory in Britain, training many graduates at Cambridge in the field.[69]

Ullmann's Cambridge career and intellectual interests differed markedly from the attitudes of his fellow émigré Geoffrey Elton, who remained markedly opposed to the pursuit of intellectual history, carrying forward a prejudice fervently enunciated by Lewis Namier, who had denounced the ideas of an age as 'flapdoodle'. The most prominent exponent of intellectual history at Cambridge, Quentin Skinner, opposed both men in articles which defended a nuanced understanding of ideas as embedded in the vocabulary of past politics. In an essay first published in 1974, Skinner had demonstrated that the eighteenth-century politician Henry St John, Viscount Bolingbroke had used ideas very firmly as a political expedient, but that the desire to be seen as a man of virtue and ideas was both prior to and indistinguishable from the expediency; in other words, ideas were of the essence of Bolingbroke's performance as a politician, and not merely ornamental 'flapdoodle'.[70] Likewise, *pace* Cowling, Skinner's refutation of Elton's narrow-minded and deliberately parochial vision of political history signalled an upwards turn in the fortunes of intellectual history in Britain.

The successes enjoyed by and the problems attendant on the Cambridge contextual school are dealt with elsewhere in this volume, but it is important to consider the achievement of the oldest associate of the school, J. G. A. Pocock, in a field separable from the history of political thought that is discussed by Richard Whatmore in this volume. Historiography was the major enthusiasm of Herbert Butterfield, and it is one shared by Pocock, his graduate pupil at Cambridge. Another enthusiasm shared by both men is study of religion as an element in political and historical thought, albeit Pocock writes about it in a sceptical and detached manner, and Butterfield wrote as a Christian apologist. A triumphant example of Pocock's reflection on such matters can be found in his masterly and now classic contribution to a *festschrift* for Butterfield which heralded a return to theology in seeking the fullest possible context in the historical exposition of political thought, in this instance through study of seventeenth-century eschatology and the religious logic of Hobbes's *Leviathan*.[71] Pocock's first book, *The Ancient Constitution and the Feudal Law* (1957) illustrated how enmeshed historiography was with political thought, and, in this sense, his work on historiography has always paid proper attention to the political and social contexts in which the works he studies emerged. This is signally true of his major work in progress, *Barbarism and Religion*, a study of the contexts and circumstances surrounding the writing and dissemination of Edward Gibbon's *History of the Decline and Fall of the Roman Empire*.[72] This is one of the major historiographical enterprises of our time, and it intersects fruitfully with his earlier classic study, *The Machiavellian Moment* (1975). It shows how study of historiography has come of age, ranging as it does across Europe and North America (and into Asia as well), drawing on history, theology, political thought, art, and literature as it does so. By showing how much Gibbon's own masterpiece is a product and reflection of the contesting cultures of the age in which it first appeared, Pocock's work gives the lie to Cowling's insistence that study of historiography, in common with the history of political thought, lacks a strong enough identity to be readily characterised as deep history. Perhaps some justice can be done to Cowling's critique, however – he insisted that historiography (like the history of political thought) 'must either become a history of the whole of thought, or present a misleading abridgement' – by noting that Pocock brilliantly abridges something very like the whole of eighteenth-century thought in his magisterial presentation of the world evoked by Gibbon in the *Decline and Fall*.[73]

Pocock has long savoured being frequently adduced as a figure in the Cambridge School despite the fact that he has long been associated with Johns Hopkins University in Baltimore. One does not need to follow his trajectory to demonstrate that the Cambridge School is an exercise in shorthand. Indeed, this can even be demonstrated from within Cambridge itself. One of the most important collections of essays to be edited within Cambridge in recent years, *Wealth and Virtue* (1983), was co-edited by Istvan Hont, a Hungarian fellow of King's who occupies the lectureship at Cambridge once held by Duncan Forbes and who had studied with Hugh Trevor-Roper at Oxford, and by Michael Ignatieff, a Canadian of Russian descent (and later to be the biographer of Isaiah Berlin), who had studied at Harvard.[74] Gareth Stedman Jones, who succeeded Quentin Skinner in the Chair of Political Science at Cambridge, had also studied at Oxford, and his turn to linguistic study in his work on class and politics in nineteenth-century Britain partly resulted from reflection on – followed by fruitful differences with – the writings of the French theoretician Louis Althusser.[75] Likewise, the philosopher closest to the Cambridge School, Raymond Guess, a fellow of King's, is an American who has spent time in Germany, whose intellectual history, including the musical legacy of the Austrian composer Alban Berg, has provided the nucleus of his research work.[76] To talk of a Cambridge School, therefore, is to beg a lot of questions, but equally importantly it is to speak of a complex entity, and not of a mere chimera.

It is remarkable that so much of the impetus for study of the Scottish Enlightenment, from Forbes to Hont and Ignatieff, has centred on Cambridge, but Scottish historians, south and north of the Border, have also importantly contributed to its study. George Davie has considered the uniqueness of the Scottish Enlightenment, whilst elsewhere in Scotland David Allan has sought to link it back to Scotland's experience of late Renaissance scholarship, and Nicholas Phillipson to its changing patterns of sociability; John Robertson, writing in Oxford, has traced its complex filiations with a comparable Enlightenment experience in Naples, and Colin Kidd, also writing in Oxford prior to a move to Glasgow University, has analysed the creation of what he calls an Anglo-British intellectual and political culture.[77] The fact that study of the Scottish Enlightenment has been pursued by scholars in England and in Scotland (and in North America) is a vivid testimony to how integrated study of intellectual history has become in Britain.

One of the foremost scholars of the development of political economy in Scotland, Donald Winch, once taught at Edinburgh University, but he has now long taught at a university which institutionalised the study of intellectual history both as an undergraduate degree and as a very individual element within what has become a department of history.[78] The University of Sussex has offered undergraduate and postgraduate degrees in intellectual history since the late 1960s (thus negating Noel Annan's lofty statement that there are no specified posts in the subject in Britain; one suspects that his work on the commission of enquiry into the student revolt at the University of Essex in the early 1970s prejudiced his views of the new 1960s universities).[79] Three of its most respected one-time members, Winch, John Burrow and Stefan

Collini, jointly produced a volume, *That Noble Science of Politics* (1983), which revolutionised study of nineteenth-century politics and political thought. Since that time, Burrow and Collini have left Sussex, Burrow – who has written illuminatingly on historiography, political thought, and on cultural as well as intellectual history – for the Chair in European Thought at Oxford, and Collini for a lectureship (since become a chair) in English at Cambridge.[80] Collini's career, which has seen him write on the origins of British sociology and on British intellectual culture more generally, is a demonstration of the strongly interdisciplinary nature of intellectual history, in that his appointment to the Cambridge English Faculty dates back to the creation of a paper on 'The English Moralists' originally associated with Basil Willey, an early member of that faculty, and himself a distinguished intellectual historian. The *festschrift* edited for Willey encapsulates in its title, *The English Mind*, the presence of the intellectual history of Britain within the study of literature.[81] The pivotal work on the language of religion and ethics in eighteenth-century England recently undertaken by the literary scholar Isabel Rivers is an eloquent testimony to the longevity and importance of Willey's legacy to students of English at Cambridge.[82]

Inevitably, perhaps, concentration on Cambridge and Oxford, the centres of most activity in intellectual history, runs the risk of looking parochial when thinking about intellectual history in Britain as a whole. Consideration naturally ought also to be given to other British universities, from Edinburgh and Glasgow to the University of Sussex outwards, but the requirements of space and the convenience for discussion provided by an analytical narrative have made this a more difficult enterprise than it might otherwise have been. What is more, a concentration of intellectual historians is only likely to be found in larger departments and faculties; it is a sub-discipline whose force is usually most deeply felt in such an environment. It is in this way, through the department of government there, that the London School of Economics, particularly through the seminar in the history of political thought associated with Michael Oakeshott, has long proved to be a major institution in the pursuit of intellectual history in Britain; its contribution to the study of intellectual history has been of signal importance.[83] Scholars who were inspired by Oakeshott at the LSE included William and Shirley Robin Letwin, American expatriates with pronouncedly British interests, and Kenneth Minogue, a New Zealander brought up in Australia; the work of J.W.N. Watkins on Hobbes demonstrated how concerned Oakeshottians were with the tone of philosophical works, bringing to the study of political thought a rather different perspective from that given it by Skinner.[84] The experience of the LSE also demonstrates how intellectual history in Britain has been largely dominated by the study of political thought.

Nevertheless, important lessons can be carried forward from the focus given in this essay. For example, at least one of its subjects, J. G. A. Pocock, is not British, but is a New Zealander who has long worked in the United

States, and whose work reflects the many complex elements of that multiple identity. He usefully complicates the notions of Britishness when considering the practice of intellectual history in Britain. Likewise, Pocock has focused on English-speaking experiences of the Enlightenment, albeit Gibbon is part of a Francophone Enlightenment as well as an English equivalent of that experience, and Pocock's pioneering status in both the 'new British history' and in studying relations between Britain and Australasia has affirmed this concentration on 'islands', including that portion of the Americas comprising the United States and Canada.[85] Where Pocock has been sceptical of embedding Britain within a monolithic understanding of Europe, Quentin Skinner has firmly identified his intellectual interests with European developments, as is most apparent in the work he has recently undertaken for the European Science Foundation.[86] Finally, reflection on Skinner's method alerts one yet further to the deeply European, frequently émigré, nature of intellectual history in Britain, as much of his approach to the study of intellectual history is indebted to the writings of Wittgenstein, being in part a product of what one might call the Wittgensteinian moment in English thought, particularly pronounced at Cambridge, and even when thinking about the trajectory of this English career we come full circle to the contested status of intellectual history in Britain. The paradox that two émigré scholars, Namier and Elton, were strongly opposed to its pursuit is more than balanced by the centrality of other émigré scholars, from Berlin and Momigliano to such contemporary scholars as Hont. The cultural understanding required of émigrés makes them great interpreters of cultures, and Britain has been deeply fortunate in acquiring such scholars: the institution of a 'Thank-Offering to Britain' lecture series by the British Academy is a signal testimony to the reciprocity of such a relationship.

Intellectual history in Britain contains many mansions, and no single approach will or can ultimately characterise it. There are signs that the history of ideas approach is flourishing again, particularly within the field of the history of political thought, and contextualism will certainly continue to flourish both in competition and in symbiosis with it.[87] Indeed, one of the major advances in the study of intellectual history in Britain has been the multiplicity of approaches taken by its practitioners. British intellectual historians have not founded a cult around any of the prominent writers in the field, and aside from the method associated with the 'Cambridge School' and the rather different 'Peterhouse School' of political-cum-intellectual history (whose continued existence beyond the writings and *obiter dicta* of the late Maurice Cowling it is hard to identify), it is satisfying to reflect that no particular style has dominated this form of historical activity. Likewise, the Oakeshottians are more concerned with detailing a disposition rather than promoting a doctrine, in keeping with the instincts of their mentor, although Perry Anderson has recently read Oakeshott as having been a much more firmly ideological thinker than such a reception would suggest.[88] Moreover, when one considers the contemporary *réclame* of one such school in the United States, that of the neo-conservatives who swarm around the thought and writings of Leo Strauss, a refugee from Nazi Germany, and of its consequent impact on social and political thinking across the Atlantic, there is room for relief that the cultivation of discipleship remains the exception rather than the rule among practitioners of intellectual history in Britain.[89]

notes

1. I am deeply grateful to Noël Sugimura and Donald Winch for generously reading and commenting on this chapter.
2. Quentin Skinner, 'The practice of history and the cult of the fact' in *Visions of Politics* (3 vols, Cambridge, 2001), I: *Regarding method*, pp. 8–26.
3. Isaiah Berlin, 'The Essence of European Romanticism' in *The Power of Ideas* (London, 2001), pp. 200–4, at p. 200.
4. Berlin, 'Russian Intellectual History' in ibid., pp. 68–78.
5. Arnaldo Momigliano, 'A Piedmontese View of the History of Ideas' in *Essays in Ancient and Modern Historiography* (Oxford, 1977), pp. 1–7, at pp. 1 and 2.
6. He was also close to another eighteenth-century historian of Namier's generation, Richard Pares; see Isaiah Berlin, 'L. B. Namier' and 'Richard Pares', in *Personal Impressions* (2nd edn, London, 1998), pp. 91–111, 120–24.
7. Isaiah Berlin, 'Nationalism: past neglect and present power' in *Against the Current: essays in the history of ideas* (London, 1979), pp. 333–55; 'The Bent Twig: on the rise of Nationalism' in *The Crooked Timber of Humanity: chapters in the history of ideas* (London, 1990), pp. 238–61; 'Jewish Slavery and Emancipation' in *The Power of Ideas*, pp. 162–85; 'Kant as an Unfamiliar Source of Nationalism', 'Rabindranath Tagore and the Consciousness of Nationality' in *The Sense of Reality: studies in ideas and their history* (London, 1996), pp. 232–48, 249–66. For a view as to how Berlin's views on nationalism compared with those of other British students of politics, see Charles King, 'Nations and Nationalism in British Political Studies' in Jack Hayward, Brian Barry and Archie Brown, eds, *The British Study of Politics in the Twentieth Century* (Oxford, 1999), pp. 313–43.
8. Isaiah Berlin, *Three Critics of the Enlightenment: Vico, Hamann, Herder* (London, 2000).
9. Isaiah Berlin, *The Age of Enlightenment* (New York, 1956); 'The Counter-Enlightenment', 'Montesquieu', 'Hume and the Sources of German Anti-Rationalism', in *Against the Current*, pp. 1–24, 130–61, 162–87; 'The Decline of Utopian Ideas in the West', 'The Apotheosis of the Romantic Will: the revolt against the myth of an ideal world' in *The Crooked Timber of Humanity*, pp. 20–48, 207–37; 'The Romantic Revolution: a crisis in the history of modern thought' in *The Sense of Reality*, pp. 168–93. For a useful discussion, see Robert Wokler 'Isaiah Berlin's Enlightenment and Counter-Enlightenment' in Joseph Mali and Robert Wokler, eds, *Isaiah Berlin's Counter-Enlightenment, Transactions of the American Philosophical Society* 93 (2003), 13–31. For spirited engagements with the themes of Berlin's explorations of Enlightenment and Counter-Enlightenment, see Aileen Kelly, 'A Revolutionary Without Fanaticism', Mark Lilla, 'Wolves and Lambs', and Steven Lukes, 'An Unfashionable Fox' in Mark Lilla, Ronald Dworkin and Robert B. Silvers, eds, *The Legacy of Isaiah Berlin* (New York, 2001), pp. 3–30, 31–42, 43–57.

10. See the exceptional essay on 'John Stuart Mill and the Ends of Life' in *Four Essays on Liberty* (Oxford, 1969), pp. 173–206. Berlin's musical tastes were also overwhelmingly European in orientation; for an instance of which, see 'The *"Naïveté"* of Verdi' in *Against the Current*, pp. 287–95.

11. Berlin 'The Three Strands in My Life' in *Personal Impressions*, pp. 255–59.

12. Berlin, 'Benjamin Disraeli, Karl Marx and the Search for Identity' in *Against the Current*, pp. 252–86; 'The Hedgehog and the Fox' in *Russian Thinkers* (London, 1978), pp. 22–81. For a suggestive discussion, see John E. Toews, 'Berlin's Marx: Enlightenment, Counter-Enlightenment, and the Historical Construction of Cultural Identities' in Mali and Wokler, *Isaiah Berlin's Counter-Enlightenment*, pp. 163–76.

13. Peter Brown, 'Arnaldo Dante Momigliano', *Proceedings of the British Academy* 74 (1988), 405–42; Oswyn Murray, 'Arnaldo Dante Momigliano', *Oxford Dictionary of National Biography* (60 vols, Oxford, 2004), 38: pp. 566–68.

14. Arnaldo Momigliano, 'Gibbon's Contribution to Historical Method' and 'George Grote and the Study of Greek History' in *Studies in Historiography* (London, 1966), pp. 40–55, 56–74; *Eighteenth-Century Prelude to Mr. Gibbon* (Geneva, 1977).

15. Arnaldo Momigliano, 'On the Pioneer Trail', *New York Review of Books* 33:8 (11 November 1976), 33–8.

16. Isaiah Berlin, 'Alleged Relativism in Eighteenth-Century European Thought' in *The Crooked Timber of Humanity*, pp. 70–90.

17. On Berlin's praise for 'negative' over 'positive' freedom, and the sense he gave to such distinctions, see 'Two Concepts of Liberty' in *Four Essays on Liberty*, pp. 119–72. More broadly, see Noël O'Sullivan, 'Visions of Freedom: the response to totalitarianism' in Hayward, et al., *The British Study of Politics*, pp. 63–88.

18. Perry Anderson, 'Components of the National Culture' in *English Questions* (London, 1992), pp. 48–104, at pp. 60–5.

19. Lewis Namier, *1848: The Revolution of the Intellectuals* (London, 1944).

20. Momigliano, *Essays in Ancient and Modern Historiography*, p. 4.

21. E. M. Butler, *The Tyranny of Greece over Germany: a study of the influence exercised by Greek art and poetry over the great German writers of the eighteenth, nineteenth and twentieth centuries* (Cambridge, 1935), pp. 3–4. She mentions Hitler as part of the process of German political mythology (at p. 333).

22. Sandra Peacock, 'Struggling with the Daimon: Eliza M. Butler on Germany and Germans', *History of European Ideas* 32 (2006), 199–215.

23. It should be noted here that a similar reaction to German Hellenism has begun to be felt among German scholars, as Mark Lilla recently noted in a trenchant review of Christian Meier's provocative study, *From Athens to Auschwitz: The Uses of History*: Mark Lilla, 'Slouching Toward Athens', *New York Review of Books* 52:11 (23 June 2005), 46–8. It is not quite fair of Lilla to refer in his opening sentence to Eliza M. Butler as 'an obscure English scholar': she may have subsequently become so, but holding the Schröder Chair in German at Cambridge was hardly a recipe for contemporary obscurity.

24. Goronwy Rees, 'A Winter in Berlin' in *Sketches in Autobiography*, ed. John Harris (Cardiff, 2001), pp. 341–64; Berlin, *Karl Marx* (London, 1938).

25. Rees, in common with Butler and Berlin, was happy to continue to generalise about national intellectual characteristics throughout his life: 'I should have remembered that what in England may seem abstractions can in Germany become the most formidable realities' ('A Winter in Berlin', pp. 359–60, 356).

26. Berlin and Wilson became friends in the late 1940s: see 'Edmund Wilson at Oxford' in *Personal Impressions*, pp. 172–82.

27. On the Institute's fortunes during this period, see Fitz Saxl, 'The History of Warburg's Library (1886–1944)' in E. H. Gombrich, *Aby Warburg: an intellectual biography* (2nd edn, London, 1986), pp. 325–38.

28. Anthony Blunt, *Nicolas Poussin* (London, 1967); Miranda Carter, *Anthony Blunt: his lives* (London, 2001).

29. Noel Annan, *Our Age: portrait of a generation* (London, 1990), pp. 224–44. For the views of a former intelligence officer and historian, see the review of Annan's book by Hugh Trevor-Roper, 'The Liberal Tide', *The Spectator* (22–29 December 1990), 57–60. I am grateful to Donald Winch for alerting me to this review, the most perceptive made at the time.

30. Noel Annan, *Changing Enemies: the defeat and regeneration of Germany* (London, 1995).

31. Noel Annan, *Leslie Stephen: the Godless Victorian* (London, 1986).

32. Annan, *Our Age*, p. 279.

33. Noel Annan, *The Dons: mentors, eccentrics and geniuses* (London, 1999), pp. 209–32.

34. See especially now the recently collected pieces that constitute *The Soviet Mind: Russian culture under Communism* (Washington, DC, 2004).

35. Isaiah Berlin, 'Joseph de Maistre and the Origins of Fascism' in *The Crooked Timber of Humanity*, pp. 91–174. For a valuable discussion, see Graeme Garrard, 'Isaiah Berlin's Joseph de Maistre' in Mali and Wokler, *Isaiah Berlin's Counter-Enlightenment*, pp. 117–31. For a telling and persuasive account of the wider context of Berlin's thought in this area, see Michael Ignatieff, *Isaiah Berlin: a life* (London, 1998), pp. 244–58.

36. Isaiah Berlin, 'The Life and Opinions of Moses Hess' in *Against the Current*, pp. 213–51.

37. Isaiah Berlin, *Freedom and its Betrayal: six enemies of human liberty* (London, 2002).

38. Isaiah Berlin, *The Roots of Romanticism* (London, 2000).

39. Arnaldo Momigliano, 'Introduction to the *Griechische Kulturgeschichte* by Jacob Burckhardt' in *Essays in Ancient and Modern Historiography*, pp. 295–305.

40. Hugh Trevor-Roper, 'The Faustian Historian: Jacob Burckhardt' in *Historical Essays* (London, 1957), pp. 273–78; 'Jacob Burckhardt', *Proceedings of the British Academy* 70 (1984), 359–78.

41. Hugh Trevor-Roper, 'Macaulay and the Glorious Revolution' in *Historical Essays*, pp. 249–53; 'The Idea of the Decline and Fall' in W. H. Barber, J. H. Brumfitt, R. A. Leigh and R. Shackleton, (eds), *The Age of the Enlightenment: studies presented to Theodore Besterman* (Edinburgh, 1967), pp. 413–30.

42. Hugh Trevor-Roper, *Archbishop Laud, 1573–1645* (London, 1940).

43. Hugh Trevor-Roper, *Renaissance Essays* (London, 1985), p. vi.

44. Hugh Trevor-Roper, 'Gibbon's Last Project' in David Womersley, ed., *Edward Gibbon: bicentenary essays* (Oxford, 1997), pp. 405–19.

45. Hugh Trevor-Roper, *Catholics, Anglicans, and Puritans: seventeenth-century essays* (London, 1987), p. vii.

46. Isaiah Berlin, 'John Petrov Plamenatz' in *Personal Impressions*, pp. 146–53; Robert Wokler, 'The Professoriate of Political Thought in England since 1914: a tale of three chairs' in Dario Castiglione and Iain Hampsher-Monk, eds, *The History of Political Thought in National Context* (Cambridge, 2001), pp. 134–58; Leszek Kolakowski, *Main Currents of Marxism* (3 vols, Oxford, 1978).

47. Berlin, *Three Critics of the Enlightenment*, p. 8. Berlin's was a classically humanistic conception of history, and of its need for the powers of imagination, on which he shared ideas with Trevor-Roper. For Berlin, see 'Historical Inevitability' in *Four Essays on Liberty*, pp. 41–117, 'The Concept of Scientific History' in *Concepts and Categories: philosophical essays* (London, 1978), pp. 103–42, and 'The Sense of Reality' in *The Sense of Reality*, pp. 1–39. For Trevor-Roper, see 'History: Professional and Lay' and 'History and Imagination', reprinted in Hugh Lloyd-Jones, Valerie Pearl and Blair Worden, eds, *History and Imagination: essays in honour of H. R. Trevor-Roper* (London, 1981), pp. 1–14, 356–69.

48. Isaiah Berlin, 'The Originality of Machiavelli' in *Against the Current*, pp. 25–79, notes at pp. 25 and 62.

49. Southern produced thoroughly contextual history in which institutional matters were always properly considered, as in his classic critique of an alleged 'school of Chartres', on which see R. W. Southern, *Scholastic Humanism and the Unification of Europe*, I: *Foundations* (Oxford, 1995), pp. 58–101. Peter Brown's early work was biographical in tone, as in *Augustine of Hippo: a biography* (London, 1967), and his later work, written in America, demonstrates how much intellectual history has tended to shade off into cultural history there, especially under the influence of Foucault, for an example of which see Brown's magisterial study *The Body in Society: men, women, and sexual renunciation in early Christianity* (New York, 1988). Brown has produced an excellent short summary of his thinking and the institutions in which it developed in *A Life of Learning: Charles Haskins Lecture for 2003* (New York, 2003). Hugh Lloyd-Jones, the Regius Professor of Greek at Oxford when Trevor-Roper was professing history there, has some claim to be a practitioner of intellectual history, as in his superb collection of essays on the history of classical scholarship, *Blood for the Ghosts: classical influences in the nineteenth and twentieth centuries* (London, 1982). A Christ Church colleague of Lloyd-Jones, William Thomas, has written about both the history of political thought and historiography, placing both within their institutional and partisan contexts: *The Philosophic Radicals: nine studies in theory and practice, 1817–1841* (Oxford, 1979), and *The Quarrel of Macaulay and Croker: politics and history in the age of reform* (Oxford, 2000). For Trevor-Roper and Laudian Peterhouse, see 'Laudianism and Political Power' in *Catholics, Anglicans, and Puritans*, pp. 40–119, and '"Little Pope Regulus": Matthew Wren, Bishop of Norwich and Ely' in H. Trevor-Roper, *From Counter-Reformation to Glorious Revolution* (London, 1992), pp. 151–71.

50. R. J. W. Evans, *Rudolf II and His World: a study in intellectual history, 1576–1612* (Oxford, 1973); *The Wechel Presses: Humanism and Calvinism in Central Europe, 1572–1627* (*Past and Present* Supplement, Oxford, 1975); *The Language of History and the History of Language* (Oxford, 1998).

51. Mark Goldie, 'J. N. Figgis and the history of political thought in Cambridge' in Richard Mason, ed., *Cambridge Minds* (Cambridge, 1994), pp. 177–92.

52. David Runciman and Magnus Ryan, 'Editors' Introduction' to F. W. Maitland, *State, Trust and Corporation* (Cambridge, 2003), pp. ix–xxix.

53. John Robertson, 'The Scottish Contribution to the Enlightenment' in Paul Wood, ed., *The Scottish Enlightenment: essays in reinterpretation* (New York, 2000), pp. 37–62.

54. Duncan Forbes, *The Liberal Anglican Idea of History* (Cambridge, 1952); 'Historimus in England', *Cambridge Journal* 4 (1951), 387–400; 'James Mill and India', *Cambridge Journal* 5 (1952), 19–33; 'Scientific Whiggism: Adam Smith and John Millar' *Cambridge Journal* 7 (1954), 643–70.

55. Duncan Forbes, 'Aesthetic Thoughts on Doing the History of Ideas', *History of European Ideas* 27 (2001), 101–13.

56. Charles Parkin, *The Moral Basis of Burke's Political Thought* (Cambridge, 1956); Eric Stokes, *The English Utilitarians and India* (Oxford, 1959).

57. Maurice Cowling, *Mill and Liberalism* (Cambridge, 1963).

58. For a perceptive critique made by another political-cum-intellectual historian, see Peter Ghosh, 'Towards the Verdict of History: Mr Cowling's doctrine' in Michael Bentley, ed., *Public and Private Doctrine: Essays in British history presented to Maurice Cowling* (Cambridge, 1993), pp. 273–321.

59. Maurice Cowling, *Religion and Public Doctrine in Modern England* (3 vols, Cambridge, 1980–2001), I, 195.

60. Ibid., i, 198.

61. Ibid., i, 433.

62. Ibid., iii, 621.

63. Ibid., iii, 619, 635–46.

64. Ibid., iii, 619–21.

65. Ibid., iii, 646–50.

66. In addition to Anderson's 1968 piece referred to in note 18 above, see Christopher Hitchens, 'Moderation or Death', *London Review of Books* 20 (26 November 1998), 3–11.

67. Quentin Skinner has also criticised, from a historical perspective, Berlin's celebrated exposition in 'Two Concepts of Liberty' of the supremacy of negative liberty: see 'The Idea of Negative Liberty: Machiavellian and modern perspectives' in *Visions of Politics*, iii: *Renaissance Virtues*, 186–212.

68. Cowling, *Religion and Public Doctrine*, i, 413. Typically, Cowling goes on to observe: 'Whether the English need to be taken out of themselves, it might be untactful to enquire.'

69. John A. Watt, 'Walter Ullmann, 1910–1983', *Proceedings of the British Academy* 74 (1988), 483–509.

70. Quentin Skinner, 'Augustan Party Politics and Renaissance Political Thought' in *Visions of Politics*, iii, 344–67.

71. J. G. A. Pocock, 'Time, History and Eschatology in the Thought of Thomas Hobbes', reprinted in *Politics, Language and Time: essays on political thought and history* (London, 1971), pp. 148–201.

72. J. G. A. Pocock, *Barbarism and Religion* (4 vols to date, Cambridge, 1999–).

73. Cowling, *Religion and Public Doctrine*, i, 230–1.

74. Istvan Hont and Michael Ignatieff, eds, *Wealth and Virtue: the shaping of political economy in the Scottish Enlightenment* (Cambridge, 1983); Ignatieff, *Isaiah Berlin*. Hont has referred happily to 'the firm wisdom' of Forbes in the acknowledgements to *Jealousy of Trade: international competition and the nation-state in historical perspective* (Cambridge, Mass., 2005), at p. x.

75. Gareth Stedman Jones, *Languages of Class: studies in English working class history, 1832–1982* (Cambridge, 1983).

76. Raymond Geuss, 'Adorno and Berg' in *Morality, Culture and History: essays on German philosophy* (Cambridge, 1999), pp. 116–39. Likewise, the Cambridge political historian Boyd Hilton, whose work has increasingly integrated intellectual history, is a product of the Oxford Modern History Faculty: see *The Age of Atonement: the influence of evangelicalism on social and economic thought, 1785–1865* (Oxford, 1988); 'The Politics of Anatomy and the Anatomy of Politics c.1825–1850' in Stefan

Collini, Richard Whatmore and Brian Young, eds, *History, Religion, and Culture: British intellectual history, 1750–1950* (Cambridge, 2000), pp. 179–97.

77. George Davie, *The Scottish Enlightenment and Other Essays* (Edinburgh, 1991), and *The Democratic Intellect: Scotland and her universities in the nineteenth century* (Edinburgh, 1961); David Allan, *Virtue, Learning and the Scottish Enlightenment: ideas of scholarship in early modern Scotland* (Edinburgh, 1993); Nicholas Phillipson, 'The Scottish Enlightenment' in Roy Porter and Mikuláš Teich, eds, *The Enlightenment in National Context* (Cambridge, 1981), pp. 19–40; 'Language, Sociability, and History: some reflections on the foundation of Adam Smith's science of man' in Stefan Collini, Richard Whatmore and Brian Young, eds, *Economy, Polity, and Society: British Intellectual History 1750–1950* (Cambridge, 2000), pp. 70–84; 'Propriety, Property and Prudence: David Hume and the defence of the Revolution' in Phillipson and Quentin Skinner, eds, *Political Discourse in Early Modern Britain* (Cambridge, 1993), pp. 302–20; John Robertson, *The Case for the Enlightenment: Scotland and Naples, 1680–1760* (Cambridge, 2005); Colin Kidd, *Subverting Scotland's Past: Scottish whig historians and the creation of an Anglo-British identity, 1689–c.1830* (Cambridge, 1993).

78. Donald Winch, *Adam Smith's Politics: an essay in historiographic revision* (Cambridge, 1978); *Riches and Poverty: an intellectual history of political economy in Britain, 1750–1834* (Cambridge, 1996).

79. Annan, *Our Age*, p. 255; *Report on the Disturbances in the University of Essex* (Colchester, 1974).

80. J. W. Burrow, *Evolution and Society: a study in Victorian social theory* (Cambridge, 1966); *A Liberal Descent: Victorian historians and the English past* (Cambridge, 1980); *Whigs and Liberals: continuity and change in political thought* (Oxford, 1988); *The Crisis of Reason: European thought, 1848–1914* (New Haven, Conn., 2000); Stefan Collini, *Liberalism and Sociology: L. T. Hobhouse and political argument in England, 1880–1914* (Cambridge, 1979); *Public Moralists: political thought and intellectual life in England, 1850–1930* (Oxford, 1991); *English Pasts: essays in history and culture* (Oxford, 1999). Collini has written about a 'Sussex School' of intellectual history in Collini et al., *History, Religion, and Culture*, pp. 1–21.

81. George Watson and Hugh Sykes Davies, eds, *The English Mind: studies in the English moralists presented to Basil Willey* (Cambridge, 1964).

82. Isabel Rivers, *Reason, Grace, and Sentiment: a study of the language of religion and ethics in England, 1660–1780* (2 vols, Cambridge, 1991–2000).

83. For a prime instance of Oakeshott's work in the history of political thought, see his *Hobbes on Civil Association* (Oxford, 1975); and for his thoughts regarding historiography, see his *What is History? and other essays*, ed. Luke O'Sullivan (Exeter, 2004). For a useful discussion of Oakeshott's approach to the history of thought, see Luke O'Sullivan, *Oakeshott on History* (Exeter, 2003), pp. 247–57. A fascinating assessment of Oakeshott's impact on both Cambridge and the LSE can be found in Annan, *Our Age*, pp. 387–401.

84. William Letwin, *The Origins of Scientific Economics: English economic thought, 1660–1776* (London, 1963); Shirley Robin Letwin, *The Pursuit of Certainty: David Hume, Jeremy Bentham, John Stuart Mill, Beatrice Webb* (Cambridge, 1963), and *The Gentleman in Trollope: individuality and moral conduct* (London, 1982); Kenneth Minogue, *The Liberal Mind* (London, 1963), and *Politics: a very short introduction* (Oxford, 1995); J. W. N. Watkins, *Hobbes's System of Ideas: a study in the political significance of political theories* (London, 1965). For an Oakeshottian critique of Skinner, see Kenneth Minogue, 'Method in Intellectual History: Skinner's

Foundations' in James Tully, ed., *Meaning and Context: Quentin Skinner and his critics* (Oxford, 1988), pp. 176–93.

85. J. G. A. Pocock, *The Discovery of Islands: essays in British history* (Cambridge, 2005).

86. J. G. A. Pocock, 'Deconstructing Europe' in Peter Gowan and Perry Anderson, eds, *The Question of Europe* (London, 1997), pp. 297–317, and 'Some Europes in their history' in Anthony Pagden, ed., *The Idea of Europe: from antiquity to the European Union* (Cambridge, 2002), pp. 55–71; Martin van Gelderen and Quentin Skinner, eds, *Republicanism: a shared European heritage* (2 vols, Cambridge, 2002); Quentin Skinner and Bo Stråth, *States and Citizens: history, theory, prospects* (Cambridge, 2003).

87. For an interesting perspective on this theme see P. J. Kelly, 'Contextual and Non-Contextual Histories of Political Thought' in Hayward et al., *The British Study of Politics*, pp. 37–62.

88. Perry Anderson, 'The Intransigent Right: Michael Oakeshott, Leo Strauss, Carl Schmitt, Friedrich von Hayek' in *Spectrum: from right to left in the world of ideas* (London, 2005), pp. 3–28. For an arguably more authentically Oakeshottian appreciation, see Robert Grant, 'Michael Oakeshott' in Mason, *Cambridge Minds*, pp. 218–37.

89. For a suggestive and thought-provoking distillation and critique of Straussian thought in America, see Anne Norton, *Leo Strauss and the Politics of American Empire* (New Haven, Conn., 2004).

3

literary and intellectual history

abigail williams

The relationship between intellectual history and literary history can be understood as reciprocal: as this chapter will demonstrate, practitioners of the history of ideas use literary texts alongside religious, scientific and philosophical writings to map the conceptual currents within an historical period, while literary critics draw on intellectual history when reconstructing the background of literary texts. However, in seeking to position texts within a history of ideas, literary critics must also confront theoretical and methodological questions about the relationship between text and context that lie at the heart of intellectual and literary history.[1]

This chapter will illustrate its arguments about the relation between literary studies and intellectual history by giving three snapshots of twentieth-century literary studies: one from the 1940s, one from the 1970s, and one from the 1980s. These three moments are chosen because each illustrates a significant shift in the literary critical understanding of historical context. As we shall see, the assertion of intellectual context and content has frequently been in conflict with questions of artistic merit and evaluation. The chapter will challenge the notion of the symbiotic relationship between philosophical and literary history by asking how far literary critics are *really* interested in the intellectual contexts of literary texts.

i

The idea of situating literary texts within their intellectual contexts is clearly not a new one. One can find myriad examples of a historicised approach to the interrelationship between literature and ideas.[2] From the 'universal history' of Polybius to the late eighteenth-century German historiography of Herder, Eichhorn, Schlosser and August Wilhelm Schlegel, numerous historians have sought to relate intellectual creations to their social environment.[3] However, the correspondence between intellectual and literary history was most systematically and famously theorised in the work of the American analytic philosopher Arthur O. Lovejoy. Lovejoy and his colleagues at Johns Hopkins University have been credited with the creation of what came to be known as the 'History of Ideas' school.

In the spring of 1933 Lovejoy gave the second series of the William James lectures at Harvard. He presented an interdisciplinary study of the ideas of plenitude, continuity and gradation in Western culture. Although Lovejoy hoped that his historical lectures would appeal to philosophers, historians and literary critics alike, the philosophers in the audience gradually dropped out, and it was the literary scholars who remained enthused.[4] In his lectures, Lovejoy identified and traced particular 'unit-ideas', as they were expressed in philosophical writing, literature, the visual arts, the sciences and social thought. The grand sweep of his narrative, which covered the period from Plato to the Romantics, was untrammelled by conventional period, disciplinary, linguistic or generic boundaries. When the lectures were published three years later, they were prefaced by an introduction in which Lovejoy outlined a formal agenda for this historical practice.[5] For Lovejoy, ideas were 'the persistent dynamic factors ... that produce effects in the history of thought'.[6] Making an analogy with analytic chemistry, he described ideas as 'component elements' of larger compound doctrines or systems. The task of the historian of ideas was both to identify and to classify those ideas, and to explore the tensions and contradictions in their use and representation across time. Indeed, Lovejoy's primary concern was classification: he writes of his intention to produce

> a study of the sacred words and phrases of a period or a movement, with a view to a clearing up of their ambiguities, a listing of their various shades of meaning, and an examination of the way in which confused associations of ideas arising from these ambiguities have influenced the development of doctrines ....[7]

Literary texts, as the prime example of the use and practice of these ideas, provided important instances of analysis. Thus in *The Great Chain of Being* Lovejoy juxtaposes literary figures with scientists and philosophers. The poet and essayist Joseph Addison is considered alongside the Dutch biologist and inventor of the microscope, Antony van Leeuwenhoek, as Lovejoy attempts to illustrate the way in which the discoveries of early microbiology exposed new worlds of natural activity and order to inspire the early eighteenth-century imagination. Likewise, rather than analysing Milton's *Paradise Lost* in the context of classical epic, or the concept of the sublime, Lovejoy discusses the poem's philosophy in relation to the nominalism of Descartes, and the writings of the Cambridge Platonists.

In seeking to promote his brand of intellectual history, Lovejoy extolled the virtues of interdisciplinary study: in the introduction to *The Great Chain of Being* he declared that one of the aims of his work was to pioneer new relations between the disciplines of philosophy and literary studies, relations which would enable historians to consider literary texts for their philosophical content, and allow literary critics to recognise the intellectual contexts of literary texts. It was an approach that inspired numerous literary scholars to investigate the evolution of specific ideas, founding the basis for the discipline of 'thematics' in literary study. Concepts as diverse as Machiavellianism, the goddess Natura, the myth of the golden age, and the notion of time all provided the subject matter for erudite intertextual studies.[8] Many of these accounts took the form of an examination of the development of the concept under discussion and, like Lovejoy's work, drew on literary and non-literary examples. Yet although Lovejoy's interdisciplinary and cross period approach clearly inspired a wealth of broad-ranging, conceptual literary history, it also exposed some of the problems inherent in assuming a direct correlation between literary and intellectual history.

In setting out the interdisciplinary basis of the history of ideas, Lovejoy claimed that

> the interest of the history of literature is largely as a record of the movement of ideas – of the ideas which have affected men's imaginations and emotions and behavior. And the ideas in serious reflective literature are, in great part, philosophical ideas in dilution.[9]

This statement seems to represent some of the difficulties implicit within Lovejoy's approach to literary study. Lovejoy effectively flattens the distinction between literary and non-literary texts, and ignores the literary texts' formal and generic aspects. Whereas contemporary literary critics sought to analyse and evaluate poems, novels and prose texts in terms of their aesthetic achievements, this, for Lovejoy, was less important than winnowing the literature of the past for its thought content. Not surprisingly, this pronouncement earned Lovejoy disapproval from a generation of literary critics trained to evaluate on formalist grounds. The marginal annotation to the excerpt quoted above in the Bodleian Library copy of *The Great Chain of Being* must have reflected the response of many contemporary readers: 'No, no, no! Literature is art, not philosophy – the chief criterion is aesthetic.'[10]

In *the Great Chain of Being*, Lovejoy uses excerpts from a series of major literary works to exemplify the development of an idea or association within a period. The form, genre, language, and context of the lines cited are not discussed, but the ideas are. Lovejoy buttresses his dismissal of aesthetic quality with additional remarks on the relationship between canonical and non-canonical writing. As he describes the characteristics of the historian of ideas, he declares that he or she is 'especially concerned with the manifestations of specific unit-ideas in the collective thought of large groups of persons, not merely in the doctrines or opinions of a small number of profound thinkers or eminent writers'.[11] This, he says, will doubtless offend students who are 'repelled when called upon to study some writer whose work *as* literature, is now dead – or at least, of extremely slight value, according to our present aesthetic and intellectual standards'.[12]

For Lovejoy, writing in an era before the theoretical assault on aesthetic essentialism, when the status of the literary canon was as yet, largely unchallenged, this apparent dismissal of literary status was to prove problematic. For in his statement about the role of literary history, Lovejoy is not quite able to uncouple value judgement from his discussion of literary texts, and so, as we have seen, he claims that 'the ideas in serious reflective literature are, in great part, philosophical ideas in dilution'. In this slightly defensive manoeuvre, Lovejoy suggests that there is likely to be a correlation between 'serious literature' and a substantial philosophical content – that there is some causal link between the two. Moreover, although he claims that the history of ideas is not confined to the work of eminent writers, his account is dominated by canonical figures – Dante, Milton, Pope, Goethe and Hugo. The notion of literary merit continues to resurface in Lovejoy's discussion of literary texts, but it is not quite clear how this merit is quantified, or what role it plays in the evaluation of the text. So, for example, commenting on a passage from Alexander Pope's *Essay on Man*, he says the lines are 'almost too familiar to quote, but too perfectly illustrative of the conception – and too superb an example of Pope's poetic style at its best – to leave unquoted'.[13] Although the history of ideas was intended to emphasise the continuity of concepts rather than matters of aesthetic evaluation, it seems that in the act of selection of literary texts, the question of poetic merit insistently raised its head. By selecting quotations from major writers, and in justifying those quotations on their literary merit, Lovejoy not only undermined his earlier commitment to the interdisciplinary and non-canonical, he also created a complex inheritance for his early followers among the literary historians. How did literary texts relate to the history of philosophy? And conversely, and perhaps more interesting for a literary historian like myself, what role had the consideration of ideas in the evaluation of a text or an author? If the question of literary value proved problematic for the philosopher, the question of the relative value and import of the conceptual content of a literary text was to test the uptake of intellectual history in literary studies. In the rest of this chapter I will challenge the notion of the symbiotic relationship between philosophical and literary history by asking whether, in fact, literary critics are really interested in the 'history of ideas'. As we shall see, the assertion of intellectual context has frequently been ancillary to the defence, interpretation and promotion of particular literary texts.

ii

We can find a very early example of the influence of the history of ideas movement in the work of Louis Bredvold. Bredvold was a member of the literature department at Johns Hopkins at the same time that Lovejoy was running his 'History of Ideas Club', and he responded to the demand frequently heard in American scholarship of the 1930s for an improved historiography of seventeenth-century literature. Like his contemporary Basil Willey, he sought to historicise the literature of the period not merely by recovering contemporary definitions, or the events behind topical references, but by identifying the broader intellectual background to a writer and their work.[14] In his account of *The Intellectual Milieu of John Dryden* (1934) Bredvold asserts that Dryden worked in a tradition of conservative scepticism, a view of the poet which was to endure, largely unquestioned, until the publication of Phillip Harth's *Contexts of Dryden's Thought* in 1968.[15] In a methodology typical of much early to mid century historicism, each chapter of *The Intellectual Milieu* takes the form of a brief overview of a subject, followed by examples of its influence in Dryden's poetry, prose and drama. Thus Bredvold explores Dryden's debts to Greek and medieval scepticism; Hobbesian materialism and the new science; fideism and Roman Catholic apologetics; and the political philosophy of Hobbes and Filmer. What is particularly interesting is that Bredvold's articulation of the Poet Laureate's intellectual context has an explicit agenda: namely, to defend Dryden's intellectual credibility. In the nineteenth and early twentieth century, Dryden's reputation as a poet suffered because he was perceived as lacking political and literary integrity.[16] By writing in celebration of both Oliver Cromwell and the Stuart monarchs, both the Anglican Church and the Catholic faith, he damned himself in the eyes of a generation of critics inclined to see political verse as the propaganda of hirelings, and a change of religious belief as mere opportunism. The recovery of the philosophical content of Dryden's works offered the opportunity to display the poet's consistency and seriousness. Bredvold says in his introduction that his study was intended to counter the assumption of Dryden as 'an expert craftsman with an uninteresting mind', engaged in ephemeral journalism whose content was riddled with inconsistency and contradiction:

> The final value of a study of Dryden's adventures among ideas lies, no doubt, in what it contributes to an understanding of the poet's personality. His versatility has long been recognised, usually with the qualifying suspicion or assumption that it precludes his having had any high purpose. But Dryden tried very hard to find himself and he eventually succeeded.[17]

In relating Dryden's characteristic inconsistencies to a European tradition of Pyrrhonism, Bredvold shores up Dryden's literary merit through claims for his intellectual rigour. One of the most revealing aspects of Bredvold's account is his acknowledgement that the ideas to which Dryden is indebted are not in themselves particularly noteworthy or original. His exploration of the intellectual context of the great writer's work concludes with the assertion that 'he was not a discoverer of new ideas': Dryden's worth can not be measured by the conceptual erudition or originality of his writing. This perception of an inverse relationship between literary and conceptual innovation was to prove a recurrent theme and problem in subsequent studies of literary texts in their intellectual context.

Theodore Spencer's *Shakespeare and the Nature of Man* (1943) offers a fruitful contrast with Bredvold's account of Dryden in that Spencer was not trying to redeem the reputation of an author who had been unjustly criticised.[18] Rather, his employment of intellectual history was intended to shed more light on the dramatist's depiction of human nature, by way of contextualisation within sixteenth-century ideas about the nature of man. Like the scholarship of his more famous, and subsequently controversial, contemporary, E. M. W. Tillyard, Spencer's book offers a reassuringly coherent 'world view' during the Second World War when such a vision of stability was particularly desired.[19] In the introduction to *Shakespeare and the Nature of Man*, Spencer establishes his difference from traditional literary criticism. He will not, he says, tell us anything about the sources of the plots, the texture of the poetry, literary fashions or dramatic devices. Instead: 'It is Shakespeare's vision of life we are after, its dependence on contemporary thought, its development through dramatic form, and its universal truth.'[20]

In tracing the 'nature of man' then, Spencer is concerned to relate the dramatisation of human existence to sixteenth-century ideas about man's place in the world. He follows a similar methodology to Bredvold, establishing first the 'context', and then reading this background into the plays. So we have an initial chapter on the optimistic theory of man in nature, offering an overview of neo-Platonism, cosmology, and the body natural and politic. This is then followed by the conflict theory, covering Burton, Goodman, Copernican theory, Renaissance humanism and Machiavelli. There follow four chapters on the plays, and the ways in which they reflect these views: so, for example, *Hamlet* is shown to allude to contemporary ideas of cosmology and kingship, while the later plays are seen to reflect 'the contemporary vision of the evil reality in man's nature'.[21] Spencer sees Shakespeare's drama, which he characterises as human experience in conflict, as produced by the tension between the two dominant opposing intellectual positions of the age: 'In the periods when great tragedy has been written, two things seem to have been necessary: first, a conventional pattern of belief and behavior, and second, an acute consciousness of how that conventional pattern can be violated'.[22]

However, although Spencer posits here a causal link between intellectual background and 'great tragedy', he is also trying to demonstrate that Shakespeare's greatness lies in his ability to transcend the historical particulars of this belief system and to communicate a 'universal truth' that is true of all human experience. So after all his discussion of Galileo and Filmer, Burton and Goodman, Spencer concludes that:

> Though he drew very largely on what he inherited of the conventional concepts of man, and though his picture of man's nature would have been very different without them, nevertheless, Shakespeare's vision of human life transcends anything given him by his time ... [He is] in Ben Jonson's familiar phrase, 'not of an age but for all time.'[23]

Like Bredvold, Spencer concludes that the recovery of the intellectual context of Shakespeare's work reveals his 'ideas' as conventional: his greatness lies elsewhere, and the ideas are ultimately used as proof of his 'literary' or aesthetic achievements, for what else could transform a series of sixteenth-century commonplaces on the nature of man into a transcendent vision of life that is 'for all time'? *Shakespeare and the Nature of Man* illustrates one of the contradictions of historical literary criticism of this period in that it attempts to mount an essentialist argument on the basis of historicist evidence. Within the transhistorical pieties of liberal humanism, intellectual history could be invoked and then dismissed as a context for artistic endeavour: in the case of Shakespeare, a consideration of sixteenth-century ideas about human nature provided evidence of all the ways in which the national bard transcended his intellectual, social, historical and political context.

iii

As we have seen, one of the functions of the early introduction of intellectual history in literary studies had been, as in the case of Louis Bredvold's Dryden, to grant writers a particular kind of intellectual credibility. The chosen author is shown to participate in a broader, more 'serious' intellectual context than previously thought. By the 1970s, this sort of intellectual defence began to seem most pressing in relation to women writers, and in particular, Jane Austen. In the criticism of the 1950s and 1960s Austen was widely acclaimed as one of England's great novelists: F. R. Leavis's *The Great Tradition* (1948) begins with the assertion 'The great English novelists are Jane Austen, George Eliot, and Joseph Conrad',[24] while in his canon-defining account of the eighteenth-century novel, Ian Watt deemed Austen the 'climax' of her genre's rise during the eighteenth century.[25] Yet Watt and others wrestled with a particularly gendered problem in constructing their canon, namely that of reconciling Austen's literary 'greatness' with the apparent 'smallness' of her feminine subject matter: stature versus scale, as Watt put it.[26] In *the Rise of the Novel* (1957), Watt discussed the 'restrictions' and the 'narrowing' associated with 'the feminine point of view', which usually outweigh the advantage of 'the greater feminine command of the area of personal relationships'. Watt's discussion bears testimony to a fundamental unease with granting high stature to a female author whose subject-matter is so unapologetically domestic and feminine, and thus apparently trivial. Austen's gender, never absent from the consideration of her work, becomes one of her most distinctive features in this critical period, and it is repeatedly associated with miniaturism.

Marilyn Butler's pioneering *Jane Austen and the War of Ideas* (1975) offered a counter to these repeated assertions of Austen's limited brilliance. Rather than praising Austen's attention to detail, Butler located the novels within the late eighteenth-century ideological struggle in England between 'sentimental' Jacobins and Tory anti-Jacobins. Arguing that all works of visual art and literature of this period were positioned within this debate, she claims that 'Jane Austen's novels belong decisively to one class of partisan novels, the conservative.'[27] Drawing on Gibbon and Burke's ideas of human nature, on traditions of sentimental latitudinarianism and anti-individualism in the anti-Jacobin literature, she explored the novels' depiction of subjective experience, emotionalism and 'naturalness'. So, for example, her rereading of *Sense and Sensibility* highlights the political ramifications of the impulsive Marianne Dashwood's insistence that feelings rather than rationality will guide her judgement. Butler concludes with a firm rebuttal of the 'small piece of ivory' school of Austen criticism that had exclusively aestheticised her achievement, arguing against the assumption that 'she does not involve herself in the events and issues of her time. The crucial action of her novels is in itself expressive of the conservative side in an active war of ideas'.[28] Yet ultimately Butler's claim for Austen's achievement is again complicated by the contrast between the writer's technical innovation and her ideological conservatism. Her ultimate defence of Austen sees her author triumphing over the limits of her intellectual content:

> Her feat is to give life to a viewpoint that, in all other hands, proves deficient in art as well as in humanity. The richness of her allusions, the intricacy with which her form embodies her ideas, have to be allowed to compensate for what is thin or partial in her presentation of the individual.[29]

Butler also raised important questions about the relationship between literary form and ideological content. Lovejoy and earlier critics had assumed a more or less straightforward relationship between text and context: namely that the literary work offered a direct reflection of its intellectual background. So, for example, Spencer's account of Shakespeare claimed that the intellectual conflict of the age provided the seedbed for a drama concerned with individuals in conflict. Yet Butler is alert to the ways in which a knowledge of the conceptual content of the work can complicate critical assumptions about what novels are and what they do. As she acknowledges in the new introduction to her 1987 edition of *Jane Austen and the War of Ideas*, she originally published the book in a literary climate in which a generation of liberal critics had declared that Austen's greatness as a novelist lay in her sympathetic depiction of the inner life. Yet what her closer examination of Austen's 'ideas' revealed was that Austen's critique of the Jacobin emphasis on selfhood, emotion and individuality effectively excluded large areas of interior experience. Butler writes:

> Her morality is preconceived and inflexible. She is firm in identifying error, and less interested than other great novelists in that type of perception for which the novel is so peculiarly well-adapted – the perception that thoroughly to understand a character is to forgive him.

Butler concludes: 'if this is true, are we right to call her a great novelist at all?'[30] The very qualities assumed to be the essence of the great novel – psychological truthfulness and forgiveness of human error – turn out to be fundamentally incompatible with Austen's intellectual heritage and commitments. What began as a historicist account of Austen's relation to the political and literary debates of the 1790s ended up interrogating the fundamentals of literary merit and the nature of the novel.

When Butler first began work on *Jane Austen and the War of Ideas*, the mere assertion of a historical context for her author and the demonstration of Austen's participation in intellectual debate was in itself a feminist act, even if it was not presented in this way. The decade after the monograph's publication saw the rise of second wave feminism in literary studies, prompting a wide scale re-evaluation of the history of women's writing and the extent to which this history reflected the oppression of female authors.[31] The historicist strain in Austen studies was now inflected far more explicitly and polemically by contemporary gender consciousness. Subsequent studies of Austen offered their own accounts of the novelist in her intellectual heritage but came to very different conclusions from Butler: Margaret Kirkham's *Jane Austen, Feminism and Fiction* (1983) established Austen's proper intellectual context as the feminist controversy of 1788 to 1810, identifying Austen as an exponent of 'Enlightenment feminism' in fiction.[32] Kirkham claims that women wrote and thought within an intellectual culture that was separate from men's, and so her background for Austen is a tradition of earlier eighteenth-century feminist writers: Mary Astell, Elizabeth Carter and Catherine Macaulay. She argues that Austen's emphasis on reason as the supreme guide to conduct is derived not from anti-Jacobin discourse, but from the rational feminism of the Enlightenment. Noting a 'striking similarity' between Austen's fiction and Mary Wollstonecraft's *A Vindication of the Rights of Woman*, she points to an underlying continuity of feminist moral concerns from *Sense and Sensibility* to *Sanditon*, which, she claims, is associated with Mary Wollstonecraft's criticism of the anti-feminist streak in Romanticism. Claudia L. Johnson's *Jane Austen: Women, Politics, and the Novel* (1988), published five years later, again engages explicitly with Butler's work, but rather than offering an account of Austen as a straightforwardly feminist writer, she says that the novelist 'interrogates' conservative doctrine and has many progressive or reformist elements in her writing.[33] Johnson argues that while Austen does not articulate the overt feminist critique of Mary Wollstonecraft and Mary Hays, her fiction, like that of Amelia Opie, Elizabeth Inchbald and Maria Edgeworth, collapses the antithetical structures of good and bad girls, and good attitudes and bad attitudes, that characterise the reactionary fictions of Edmund Burke and his conservative adherents. She concludes that Austen shares with her more politicised contemporaries a 'commitment to uncovering the ideological underpinnings of cultural myths'.[34]

Johnson not only takes issue with Butler's conclusions about Austen's conservatism, but also Butler's literary-historical methodology, and in particular her treatment of Austen's intellectual context.

> She [Butler] contends at last not that her [Austen's] ideas were engaged and developed by that very war [of ideas], but rather that 'old fashioned notions' were 'given to her' by the 'sermons' and 'conduct books' that somehow 'formed' her mind. This account, however, actually denies Austen the dignity and the activity of being a warrior of ideas.[35]

For Johnson, then, the assumption that Austen merely inherited a set of conservative ideas which she duly reflected in her novels is a model of text and context that denies the female author agency. Johnson's account of the subversive Austen sees her instead shaping her intellectual climate, offering a series of dynamic interventions in contemporary political discourse.

This snapshot of Austen studies reveals just how far the feminist debate of the 1970s and 1980s both questioned the type of intellectual history used to understand Austen's fiction, and also challenged assumptions about the way in which the writer related to that context. We can see how the notion of 'intellectual background' as a stable, and objectively reliable contextual framework for the literary text was becoming increasingly unsustainable in the heat of the political and theoretical debates of the later 1970s and 1980s. Jane Austen's 'ideas' and 'intellectual context' could either be the conservative Christian morality of the 1790s, or an early feminist critique of Burkean patriarchy, depending on the theoretical positioning of the critic. Moreover, it was no longer acceptable merely to note the influences on a writer's work: the model of the author as passive mouthpiece of contemporary intellectual debates was refashioned to emphasise the proactive and transformative power of the woman writer.

iv

Theodore Spencer had opened his account of *Shakespeare and the Nature of Man* with the statement that 'the first thing that impresses anyone living in the twentieth century who tries to put himself into Shakespeare's intellectual background is the remarkable unanimity with which all serious thinkers, at least on the popular level, express themselves about man's nature and his place in the world.'[36] As he and other immediately pre- and post-war critics, most famously Tillyard, author of *The Elizabethan World Picture* (1943), saw it, though the modern era might be a time of social, political and intellectual fragmentation, the early modern period provided a halcyon era of homogeneity and shared values.[37] As we can see, by the time of the publication of Claudia Johnson and Marilyn Butler's work on Jane Austen, it had become clear that there could be multiple and conflicting versions of intellectual history within which to situate a literary text. Yet these texts were still predicated on the assumption that the 'correct' intellectual context could be found for Austen's work. The New Historicism of the late 1980s and 1990s set out to challenge some of the epistemological certainties about historical context that had characterised earlier literary criticism. Stephen Greenblatt, Louis Montrose and Catherine Gallagher and their followers sought to find ways of readdressing the relationship between text and historical context. They did so in the light of an anthropological and post-structuralist understanding of culture in which the notion of a single worldview or historical context had come to seem increasingly untenable, and in which art and society had come to be seen as related institutional practices.

Although the leading practitioners of new historicism have continued to resist any programmatic definition or manifesto for their approach, they shared a commitment to the exploration of the variety of ways that early modern texts were situated within the larger spectrum of discourse and practices that constituted sixteenth- and seventeenth-century English culture.[38] Drawing on extra-literary matters, including letters, diaries, films, paintings and medical treatises, they sought to reposition literary texts in a dynamic and often oppositional relation to their culture, seeking 'surprising coincidences' that could cross generic, historical, and cultural lines. Where, as we have seen, Butler and Johnson offered, respectively, accounts of Austen that either reflected *or* interrogated contemporary ideology, the New Historicists embraced the possibility that a literary text might do both at the same time. The literary text bears a dialectical relationship towards its 'context' in that it can both represent a society's behaviour patterns, and perpetuate, shape or alter that culture's dominant codes. In the introduction to his influential *Renaissance Self-Fashioning: From More to Shakespeare* (1980), Greenblatt declares his intention to abandon a traditional historicist distinction between text and content, arguing that

> if literature is seen only as a detached reflection upon the prevailing behavioural codes, a view from a safe distance, we drastically diminish our grasp of art's concrete functions in relation to individuals and to institutions, both of which shrink into an obligatory 'historical background', that adds little to our understanding.[39]

Instead, he declares, he seeks to understand literature 'as part of a system of signs that constitutes a given culture; its proper goal ... is a poetics of culture'.[40] Greenblatt positions his new historicist approach as a riposte to, on the one hand, the hermetically sealed aestheticism of the New Criticism, and, on the other, to the 'worldview' model of historical background which represented the past as a static and homogeneous backdrop to works of genius.

Renaissance Self-Fashioning was published well over two decades ago, and recent years have seen the waning of the new historicist moment, its former hegemony in early modern literary studies replaced by manuscript studies, cultural studies and the history of the book. Yet it remains the case that the New Historicism, and its theory that literary creations are cultural formations shaped by 'the circulation of social energy', has done more to re-energise debate about historicism, literary value and the relationship between text and context in the past 25 years than any other comparable theoretical approach – and as such, offers a fit conclusion to this analysis of the use of intellectual history in literary criticism. On the surface, the New Historicism sought to address some of the problems we have seen in earlier historicist criticism: namely of the separation of literature and context; and the assertion of the literary text's 'transcendence' of its time and place. In seeking to challenge the grand narratives of the 'worldview' school of intellectual history, the New Historicists characteristically drew on smaller narrative units – anecdotes – in order to illustrate the contingency and particularity of the historical moment. These anecdotes offered selective heuristic moments within which one might discern the complex intersection of literary, social, economic, intellectual and political discourses. Thus Greenblatt's account of More in *Renaissance Self-Fashioning* famously begins with an account of a dinner party at Cardinal Wolsey's house in 1534. The fictional reconstruction of the occasion is used as the basis for a range of claims: to suggest More's humanist indignation at clerical abuses; the communion and the vanity of human society, and his observation of social comedy, all of which Greenblatt sees as central to More's subsequent self-fashioning and self-cancellation. It is a brilliant and arresting opening and offers a moment in which we momentarily lose the mechanical division between the author and his background that characterised traditional historicist criticism. As Greenblatt has subsequently written, the anecdote 'offered access to the everyday, the place where things are actually done, the sphere of practice that even in its most awkward and inept articulations makes a claim on the truth that is denied to the most eloquent of literary texts'.[41]

One of the most pervasive criticisms of this methodology has been that the new historicist's extrapolation of large epistemological, historical and political claims from a single text or anecdote hangs a large weight on a small nail. But another consequence of this use of anecdote, demonstrated in Greenblatt's essay on More, is that in giving so much attention to the surprising, to the rare moments of communion with the dead, the critic is left little room to account for the important but *unsurprising* aspects of More and his work.

Greenblatt's account blends a reading of the Utopia with the dinner party, a discussion of *The Ambassadors*, a description of the theatricality of dress at the court of Henry VIII, and biographical narrative, but in trying to capture moments of More's self-fashioning through these unlikely juxtapositions, he is less able to account for the specifics of More's intellectual self-fashioning. There is no substantial discussion of More in the broader setting of Northern European humanism; of the educational practices and theories that fostered the rhetoric and argument of More's work; of the role of Lucianic argument and dialogue. Greenblatt certainly intends to read More for and in the bigger picture, identifying in his writing a 'system of signs that constitutes a given culture', but the systems and the signs that he finds are those largely derived from the social sciences and new cultural history that so influenced New Historicist methodology. Thus we see More in relation to ideas of guilt, honour, patrimony – but not what he might have read, how he might have been taught, who he might have corresponded with, how far his ideas were derived from medieval sources. And it is interesting to see that Greenblatt's assertion of More's participation in contemporary ideas of, for example, guilt or honour, is based on assumed familiarity with the 'thought of the age' rather than direct evidence. In this the New Historicist resembles the old historicism typified by Spencer, who emphasised that Shakespeare might well not have read the authors that he was citing as representative of the sixteenth-century debate over the nature of man.[42] Greenblatt's More reflects an early modern understanding of confession, or shame, in much the same way as Spencer's Shakespeare reflected his age's familiarity with cosmology or scepticism. As we have seen, New Historicism was born out of a desire to avoid grand narratives, and to avoid 'worldview' statements about the thought of an age. Yet it seems that it is ultimately impossible to undertake literary-historical criticism without reducing some of that history to 'background'. While there is a strong sense of the complex and intangible nature of history and historical event in Greenblatt's essays, there is also historical 'background' constructed through a series of references to secondary accounts of the period. One of the consequences of the interdisciplinary eclecticism of New Historicism is that it demands of the critic a vast range of expertise. In order to read correctly all the signs of a culture, the literary critic must turn social historian, anthropologist, philosopher, art historian and, inevitably, has to rely on other authorities to support those readings. Thus the contingency of historical event tends to harden into background fact. So Greenblatt alludes to 'a belief in inherited family characteristics, such as was widespread in the Renaissance',[43] to the planning of patrimonial inheritance as 'according to Natalie Davis one of its [the age's] central and defining concerns'.[44] He declares that, 'as the public, civic world made increasing claims on men's lives, so, correspondingly, men turned in upon themselves, sought privacy',[45] a claim substantiated with a single reference to an unpublished essay. While all historiography is necessarily dependent upon secondary sources, there is in the New Historicism a tension between the 'touch of the real' as established in anecdote and juxtaposition, and the hegemonic and static nature of the authoritative historiography recruited in support of its broader narratives.

Greenblatt, like the earlier critics discussed above, wrestles with the relationship between the writer and the world they may or not represent. New Historicism is often equated with the dismantling of the canon: with the idea that, for example, Shakespearean tragedy can be read on a par with a contemporary pamphlet. But this is not actually true of what Greenblatt is doing in his seminal work: *Renaissance Self-Fashioning* is a collection of essays about canonical authors of the sixteenth and seventeenth century, and it is focused on their major works. In the introduction, Greenblatt justifies his selection:

> from the thousands, we seize upon a handful of arresting figures who seem to contain within themselves much of what we need, who both reward intense, individual attention and promise access to larger cultural patterns.
>
> That they do so is not, I think, entirely our own critical invention ... We respond to a quality, even a willed or partially willed quality, in the figures themselves, who are, we assume by analogy to ourselves, engaged in their own acts of selection and shaping and who seem to drive themselves toward the most sensitive regions of their culture, to express, and even, by design, to embody its dominant satisfactions and anxieties. Among artists the will to be the culture's voice – to create the abstract and brief chronicles of the time – is a commonplace, but the same will may extend beyond art.[46]

We have here a reversal of the methodology of Old Historicism: in the works of Bredvold, Spencer and Butler we had a series of chapters on the ideas of a period, which were then read into the literary texts. In this New Historicist model we have the text which then reveals the world: Thomas More tells us about opposition to and complicity with the Elizabethan state; Edmund Spenser tells us about colonial identity in sixteenth-century Ireland. Greenblatt claims that we can learn about the age from the figure who will 'promise access to larger cultural patterns', and in this book, he makes arguments about the construction of selfhood in the early modern period from the evidence of More, Tyndale, Wyatt, Spenser, Marlowe and Shakespeare. Rather than transcending their age, as they would have done 40 years previously, these writers now *are* their age – a claim which is no less problematic.

In the passage cited above, Greenblatt is curiously silent on the subject of the literary merit of his selected writers: he asserts the historiographical usefulness of the literary text as an index of the energies of a culture, and claims that writers 'reward intense, individual attention' but seems reluctant to broach the matter of their literary merit, of the power that is particular to literature.[47] It is a position that seems to take us back to Lovejoy's dictum that '... the interest of the history of literature is largely as a record of the movement of ideas'. In both cases, the literary work is seen in essentially utilitarian terms: for what it can tell us about the history of an idea, or about the 'cultural practices' of a particular period. And conversely, when we juxtapose this with the historicism exemplified in the work of Bredvold, Spencer, Butler and Johnson, we find a tradition of literary criticism in which the history of ideas has served as a flexible backdrop to the defence of an author's seriousness, their radicalism or their ability to transcend time – but rarely as an area of interest in its own right. I have, of course, here overemphasized the oppositional nature of these approaches, and also oversimplified the opposition between literary formalism and literary historicism, since, not least, any particular instance of criticism almost always draws on both text and context to some degree. Yet it is clear from this brief survey alone that the possibility of accommodating intellectual history within literary criticism foregrounds a series of broader questions about the nature of text in context: what is the relationship between literary and conceptual innovation? Can a literary work transcend its intellectual context? Is the literary text a reflection or an agent of intellectual debate? Can it really be both at the same time? What role has intellectual history alongside the various economic, social, biographical contexts that we might use to explore a text in its historical moment? They are all issues that will continue to define debate over the role of historicism in literary studies.

notes

1. For surveys of the relationship between literary criticism and the history of ideas, see Timothy Bahti, 'Literary Criticism and the History of Ideas' in Christa Knellwolf and Christopher Norris, eds, *The Cambridge History of Literary Criticism*, Vol. 9, *Twentieth-Century Historical, Philosophical, and Psychological Perspectives* (Cambridge, 2001), pp. 31–42; Richard Macksey, 'History of Ideas' in Michael Groden, Martin Kreiswirth and Imre Suzman, eds, *The Johns Hopkins Guide to Literary Theory and Criticism* (2nd edn, Baltimore, Md, and London, 2005), pp. 499–504.

2. For a brief summary of the 'History of Ideas' before Lovejoy, see Donald R. Kelley, 'What is Happening to the History of Ideas?', *Journal of the History of Ideas*, 51 (1990), 1–11.

3. On the German historians of culture, see Wilhelm Dilthey, 'Das achtzehnte Jahrhundert und die geschichtliche Welt' (1901), in Dilthey, *Gesammelte Schriften III* (Leipzig, 1927), pp. 209–68; Friedrich Meinecke, *Die Entstehung des Historismus* (2 Vols, Frankfurt am Main, 1936); Heinz Schlaffer and Hannelore Schlaffer, *Studien zum ästhetischen Historismus* (1975).

4. Daniel J. Wilson, *Arthur O. Lovejoy and the Quest for Intelligibility* (Chapel Hill, NC, 1980), p. 191.

5. Arthur O. Lovejoy, 'The Historiography of Ideas', *Essays in the History of Ideas* (Baltimore, 1948).

6. Arthur O. Lovejoy, *The Great Chain of Being: A Study of the History of an Idea* (Cambridge, Mass., 1936), p. 5.

7. Ibid., p. 14.

8. See J. G. A. Pocock, *The Machiavellian Moment: Florentine political thought and the Atlantic republican tradition* (Princeton, NJ, 1975); George D. Economou, *The Goddess*

Natura in Medieval Literature (Cambridge, Mass., 1974); Harry Levin, *The Myth of the Golden Age in the Renaissance* (Bloomington, Ind., 1969); Ricardo J. Quinones, *The Renaissance Discovery of Time* (Cambridge, Mass., 1972).

9. Lovejoy, *The Great Chain of Being*, pp. 16–17.
10. Ibid., p. 16. Bodleian Library shelfmark, URR A.1.285.
11. Ibid., p. 19.
12. Ibid.
13. Ibid., p. 199.
14. See Basil Willey, *The Seventeenth Century Background : studies in the thought of the age in relation to poetry and religion* (London, 1934).
15. Louis Bredvold, *The Intellectual Milieu of John Dryden* (Ann Arbor, 1934); Phillip Harth, *Contexts of Dryden's Thought* (Chicago, Ill., and London, 1968).
16. For an overview of this critical inheritance see David J. Latt and Samuel Holt Monk, *John Dryden: a survey and bibliography of critical studies, 1895–1974* (Minneapolis, Minn., 1976), pp. 3–7.
17. Bredvold, *The Intellectual Milieu of John Dryden*, p. 152.
18. Theodore Spencer, *Shakespeare and the Nature of Man* (Cambridge, 1943).
19. For a fuller discussion of Spencer and Tillyard in the context of twentieth-century historical criticism of Shakespeare, see Michael Taylor, *Shakespeare Criticism in the Twentieth Century* (Oxford, 2001), pp. 168–72.
20. Spencer, *Shakespeare and the Nature of Man*, p. viii.
21. Ibid., p. 123.
22. Ibid., p. 50.
23. Ibid., p. 208.
24. F. R. Leavis, *The Great Tradition: George Eliot, Henry James, Joseph Conrad* (London, 1948), p. 1.
25. Ian Watt, *The Rise of the Novel: studies in Defoe, Richardson and Fielding* (Berkeley, Calif., 1957). For a full survey of Austen criticism in this period, see Eleanor Collins, 'Reading Gender, Choice and Austen Narrative' (unpublished doctoral thesis, University of Oxford, 2005), chapter 2.
26. See especially Ian Watt, 'Introduction' in *Jane Austen: A Collection of Critical Views* (Englewood Cliffs, NJ, 1986), pp. 2–3.
27. Marilyn Butler, *Jane Austen and the War of Ideas* (Oxford, 1975), p. 3.
28. Ibid., p. 294.
29. Ibid., p. 298.
30. Ibid.
31. Seminal works of this period include: Patricia Meyer Spacks, *The Female Imagination: a literary and psychological investigation of women's writing* (London, 1976); Sandra M. Gilbert and Susan Gilbar, *The Madwoman in the Attic: the woman writer and the nineteenth-century literary imagination* (New Haven, Conn., and London, 1979).
32. Margaret Kirkham, *Jane Austen, Feminism and Fiction* (Brighton, 1983).
33. Claudia L. Johnson, *Jane Austen: women, politics, and the novel* (Chicago and London, 1988).
34. Ibid., p. 27.
35. Ibid., pp. xvii–xviii.
36. Spencer, *Shakespeare and the Nature of Man*, p. 1.
37. E. M. W. Tillyard, *The Elizabethan World Picture* (London, 1943).
38. For a discussion of some of the origins and objectives of the first New Historicists, see Catherine Gallagher and Stephen Greenblatt, *Practicing New Historicism* (Chicago, Ill., and London, 2000), pp. 1–19.

39. Stephen Greenblatt, *Renaissance Self-Fashioning: from More to Shakespeare* (Chicago and London, 1980), p. 4.
40. Ibid., pp. 4–5.
41. Stephen Greenblatt, 'The Touch of the Real' in *Practicing New Historicism*, pp. 20–48.
42. Greenblatt, *Renaissance Self-Fashioning*, p. x.
43. Ibid., p. 43.
44. Ibid.
45. Ibid., p. 46.
46. Ibid., pp. 6–7.
47. Stephen Greenblatt's most recent work has signalled a renewed interest in literary power, and of all the elements of the literary text that cannot be explained by its role within cultural systems. See, in particular, *Hamlet in Purgatory* (Princeton, 2001) and *Will in the World: how Shakespeare became Shakespeare* (London, 2004).

4

intellectual history and the history of art

lucy hartley

Let us search more and more into the Past; let all men explore it, as the true fountain of knowledge; by whose light alone, consciously or unconsciously employed, can the Present and the Future be interpreted or guessed at. For though the whole meaning lies far beyond our ken; yet in that complex Manuscript, covered over with formless inextricably-entangled unknown characters, — nay which is a *Palimpsest*, and had once prophetic writing, still dimly legible there, — some letters, some words, may be deciphered; and if no complete Philosophy, here and there an intelligible precept, available in practice, be gathered: well understanding, in the mean while, that it is only a little portion we have deciphered; that much still remains to be interpreted; that History is a real Prophetic Manuscript, and can be fully interpreted by no man.

Thomas Carlyle, 'On History' (1830)[1]

The activity of an historian, as described by Thomas Carlyle, is a difficult and demanding endeavour, designed to produce truths about past worlds that will enable us to interpret present and future conditions, and yet destined to be incomplete and incoherent because the manuscript of history is ultimately beyond interpretation. This double vision of history as both an illumination, 'the true fountain of knowledge', and a palimpsest, of 'formless inextricably-entangled unknown characters', is highly significant because it renders translation of the historical record necessary for the advancement of knowledge while also disavowing the possibility that any sort of interpretive act will be sufficient to understand the prophetic writing that constitutes its invisible core. Conceived in this way, history involves two different models of temporality with the recording of the past via the manuscript differentiated from the overwriting of the past in the palimpsest: the former suggests a sense of history as story, telling of successive events within certain conventions of style, whereas the latter indicates a view of history as conversation, with competing versions of the same tale overlaying, and perhaps erasing, one another. As the broader context for the above passage makes clear, Carlyle was preoccupied with the disjunction between the manuscript account of history as 'successive events' and the palimpsest version as an 'aggregate of activities':

> as all Action is ... to be figured as extended in breadth and in depth, as well as in length ... so all Narrative is ... of only one dimension; only travels forward towards one, or towards successive points: Narrative is *linear*, Action is *solid*.[2]

What emerges most sharply in Carlyle's vision of history is the danger of smoothing out the many and varied occurrences of the past into simple narratives of rise and fall or triumph and failure. Thus, replacing manuscript with palimpsest in pursuit of prophecy, he perceived history as complex and, to a large degree, elusive: it was a repository of knowledge, containing precepts and practices that might be intelligible to the skilled observer but, in all likelihood, might not provide more than a partial view of the vast subject named 'History'. To put it in other words, Carlyle linked the purpose of history to the discovery of a mystical sort of knowledge, indexing history to prophecy as the absolute source of truth – an authority intuited, not learned, and so existing outside the everyday course of things and beyond the reach of the ordinary observer.[3] Yet also emphasised was the fundamental weakness of presenting a specific kind of intellectual enquiry as a singular or one-sided pursuit, which implied the strengths of an approach to history as pre-eminently pluralistic – the interpretation of distant lives, peoples and actions that encroach upon and help to define one another.

The claims and contradictions tied up in this understanding of history are not, of course, unique to Carlyle.[4] Arguments for the continuities between past, present and future, the coherence of the historical record (as something like a key to all mythologies), and the narration of historical events (as an inexorable progress towards a universal truth) have all been declared and delineated as well as disputed and debunked in discussions of how we understand the past that usually occur in the academic discipline of history.[5] Disciplines – subjects of knowledge or scholarly fields of enquiry – are complex entities in that the formation of a discipline can be understood as an expansive means of clarification or a limiting form of regulation for a specific type of knowledge.[6] Most of the disciplines that are established in universities today have their origins in a gradual ossification of tacit distinctions between kinds of knowledge and modes of intellectual enquiry.[7] From the middle third of the nineteenth century, what might be known about the world, about nature and culture as well as man and other kinds of animals, began to be sectioned out into distinct fields of study and substantial gaps appeared initially between the natural sciences (geology, biology, botany and chemistry) and the social sciences (politics, economics and psychology), and later between these areas and the humanities (philosophy, history, literature and art). Consequently, disciplines today appear to have very different sets of assumptions about their methodologies and critical practices, and the boundaries between them seem to be arbitrary but fixed constructions, designed to serve the purposes of institutions such as universities, libraries and archives in classifying knowledge.[8] Why is it necessary to categorise intellectual activity into discrete disciplines? What do we gain from a rendering of history that focuses on a singular aspect of human activity? And, conversely, what do we lose from our picture of human nature as a result of this process of atomisation?

   This chapter considers these questions in the light of a specific instance of disciplinary formation in the humanities, which seems, at the same time, to resist disciplinary categorisation. The history of art has many different forms and involves a range of different spheres of activity, from the connoisseurship of the collector in the auction house and salesroom, and the studio practice of the professional artist, to the classifications of the curator in museums and galleries, and the teachings of an academic art historian. However, I do not attempt to trace the development of the history of art as a discipline or to isolate new artistic forms and movements in distinct historical periods; nor do I intend to identify either a canonical list of authors who have contributed to the growth of such a discipline or a critical vocabulary that can be claimed as specific to the art-historical discipline.[9] Instead, I consider the relationship between two closely connected kinds of history, the history of art and intellectual history, which developed alongside each other in the nineteenth century as modes of enquiry into political thought, artistic production and cultural life.[10] As Stefan Collini, Donald Winch and John Burrow have recently contended:

> The defining vocation of the intellectual historian is to be alive to the several dimensions of the thought and feeling of the past ... to attend to the sensibilities of our authors, to the emotional and aesthetic satisfactions they derived from their views, to the styles and genres in which they chose (and in some cases could not but choose) to express them, and to the constant tensions between, on the one hand, the unevenly experienced pressure towards coherence, and, on the other, the desire, not always articulated or even acknowledged, to hold on to certain deeply felt intuitions, a process of internal negotiation which was itself always subject to the demands of occupying a particular station or of addressing a particular audience.[11]

This is, obviously, no small task. Both intellectual history and the history of art appear to share the goal of making sense of the past in order to offer a 'thickly textured characterization' of a specific cultural moment.[12] I want, therefore, to show how the close, even intimate, alignment of intellectual history and the history of art sets a certain style for writing about the past, the virtues of which are inextricably linked to the survival of humanism as an educative goal.

The chapter begins by outlining two lines of thought concerned with the question of how culture is constituted and developed via works of art. The first part identifies the themes, and examines the ideas of states, cultures and personalities advanced by Jacob Burckhardt, while the second part explores the contributions of John Ruskin and Walter Pater to discussions of the purpose of art in a modern, industrialised world. The third part shows how these lines of thought converge in the intellectual project of the Warburg (and later Courtauld) Institute and *Journal* from the 1930s onwards, with specific reference to the work of Aby Warburg and Ernst Gombrich. And the final part looks at current thinking about the relation of art and history to culture, in particular practitioners of work that marries intellectual and art historical interests such as Michael Baxandall, John Barrell and Ludmilla Jordanova. What kinds of histories do these authors formulate? Do they extend the enquiries traditionally associated with intellectual history and the history of art? And to what extent (and in what ways) do they challenge them? These are the concluding questions the chapter will address.

i

Jacob Burckhardt's *The Civilization of the Renaissance in Italy*, first published in 1860, had an initial print run of 1,000 copies and was not an instant bestseller. Burckhardt (1818–1897), Professor of History at Basle University in Switzerland, wanted to produce a chronicle of the cultural history of man, from the time of Constantine the Great in the fourth century AD to the Renaissance of the fourteenth and fifteenth centuries, which would chart the gradual transitions from Pagan to Christian and then to humanist worldviews.[13] Yet only the first and last volumes in the planned series were completed: *The Age of Constantine the Great* (1853) told of the decline of ancient civilisation and the emergence of a new culture in the West, while *The Civilization of the Renaissance* (1860) sketched the demise of the Middle Ages and the rise and triumph of the Renaissance, identified as the beginning of modern times.[14] A second edition of *The Civilization of the Renaissance* (1868) was translated into English by S. G. C. Middlemore in 1878, but it took more than 20 years after the release of a third edition (1878) for sales to start reflecting the profound importance of a work that forged vital, and powerful, connections between culture, art and politics.[15]

Burckhardt's secular historiography of the Renaissance, from the birth of Dante to the death of Michelangelo, was organised into six parts: the state as a work of art; the development of the individual; the revival of antiquity; the discovery of the world and of man; society and festivals; and morality and religion. Together, these elements contributed to a densely wrought picture of Renaissance thought and feeling, constituted in state structures and individual groups, and performed via debates about political goals and personal interests. The result, in the broadest sense, was a book that indicated the ways and means through which the life of the past, seen through the activity of its peoples, might enable a deeper understanding of the complexities of modern society in nineteenth-century Europe.[16] Thus, Burckhardt's conceptualisation of the Renaissance depended on narrating the historical specificity of a specific span of time, and arguing for its continuing significance as 'a civilization which is the mother of our own, and whose influence is still at work among us'.[17] This extension of the frame of reference, from an historical then of the fourteenth and fifteenth centuries to an historical now in the nineteenth century, was a defining feature of Burckhardt's method; and yet, as he explained, the potential weakness of this approach was that 'a great intellectual process must be broken up into single, and often into what seem arbitrary categories, in order to be in any way intelligible'.[18] In stressing the need to find a criteria of selection that would convey adequately the breadth and depth of his material, Burckhardt acknowledged the twin difficulties of perspective and bias; each individual, he said, has their own way of seeing the materials of the past and so will develop very different kinds of emotional and judgemental responses to a given civilisation. And that, indeed, was the point.

Drawing historical and theoretical parallels from Renaissance to modern times, Burckhardt sought to make a virtue out of the diversity of opinions and multiplicity of arguments captured and communicated in his study. He examined a range of political activities relating to the legitimacy of governance, such as, for example, the rule of Emperor Frederick II over Lower Italy and Sicily, and the hold of the *Condottieri* on the balance of power; importantly, the *Condottieri* were bands of mercenary soldiers who participated in local wars between rulers of Italian states, often establishing themselves as independent rulers and including such intriguing men as Francesco Sforza and Federigo da Montefeltro, Duke of Urbino amongst their number. Likewise, he considered the cultural ramifications of religion, art and education on the development of the state, including the domination of the papacy over the people, especially (and sometimes dubiously) Nicholas V, Innocent VIII, Julius II and Leo X; the aestheticisation of life under the Medici reigns of Lorenzo the Magnificent and Cosimo de Medici; and the educative role of the humanists, specifically Petrarch, Dante, Boccaccio and Machiavelli. Throughout, Burckhardt's goal was not to smooth out tensions and contradictions but to stress the diversity of sensibilities, views, styles and genres of these many and varied examples, for the underlying thesis was that the intersection of political and cultural forces in Renaissance Italy produced an epoch-making cultivation of personality.

To illustrate this argument, Burckhardt configured the narrative as a war between the opposing forces of Christianity and humanism, with both camps expressing a desire for political rule in Italy but each offering very different opinions about how that rule should be constituted, that is to say on what grounds and to what ends. He contended that the rejection of feudalism in fourteenth-century Italy had the effect of placing the power to govern states in an extremely precarious position between distinct types of political structures, each embodying very different opinions about the advantages (or disadvantages) of national unity. The emperors and the papacy, representing the old order of things, were challenged by the newer political units of republics and despots, with the result that the new republicans supplied the catalyst for the first stirrings of modernity:

> In them for the first time we detect the modern political spirit of Europe, surrendered freely to its own instincts, often displaying the worst features of an unbridled egotism, outraging every right, and killing every germ of a healthier culture. But, wherever this vicious tendency is overcome or in any way compensated, a new fact appears in history – the State as the outcome of reflection and calculation, the State as work of art.[19]

Modern, state, reflection, calculation, art: these were the keywords in Burckhardt's account of the construction of a new kind of human consciousness, a consciousness derived from comprehension of the intricate ties connecting self to society and state, and fostered out of an individualistic creed of egotism, rights and self-interest.

A number of elements were crucial in creating the right conditions for this (re)birth of human consciousness into reflection and calculation; in particular, the appeal to the old Roman Empire in the imitation of antiquity, and the recognition of the profound importance of Venice and Florence as models of independent republics. And yet, paradoxically, the dissemination of a secular understanding of individual man (primarily in the promotion of a humanistic philosophy and principally via literature and the arts) represented both the necessary precondition for, and a potentially powerful form of resistance to the emergence of the 'state as a work of art'. Above all, then, Burckhardt narrated a story about the development of the human race as fashioned from the rise and fall of political regimes founded on envy, ambition, conspiracy, idealism, corruption and murder. Venice, 'the city of apparent stagnation and of political secrecy', recognised its own identity from inception 'as a strange and mysterious creation – the fruit of a higher power than human ingenuity'.[20] Largely because of its commercial operations, harnessed to a firm understanding of statistical science, the Venetians were bound together in mutual economic interest: 'the supreme objects were the enjoyment of life and power, the increase of inherited advantages, the creation of the most lucrative forms of industry, and the opening of new channels for commerce'.[21] As Burckhardt made clear, Venice may have lacked in literary accomplishment but it achieved ample compensation in its status as the first representative of the economic aspect of modern political life. Nonetheless, to his mind, Florence, 'the city of incessant movement',[22] deserved the greatest praise and admiration because it fostered intellectual freedom and independence, uniting the 'most elevated political thought' and the 'most valued forms of human development' in order to bring into being what Burckhardt rather grandly described as the 'first modern State in the world'.[23] Florence was superior to any other Italian city, including Rome and Venice, because it acted like a 'living organism', capable of perceiving its own civic development and progression as 'a natural and individual process'.[24]

The application of an organic (or transformative) metaphor of growth to the description of ideal statehood, whereby physical variability and material change were seen as part of a process of adaption to environment, constituted only half of the historiographical paradigm advanced by Burckhardt in *The Civilization of the Renaissance*, however. The other half of this paradigm, less concerned with general categories than with particular experiences, relied on the mystical metaphor of a veil to show how human consciousness, concealed in the moral order of the Middle Ages, was revealed in the secular humanism of Renaissance Italy. The point was that humanism disclosed the possibility of intellectual emancipation in so far as it affirmed the value of individual identity and, crucially, the ideological power of self-representation. As Burckhardt explained:

> In the Middle Ages both sides of human consciousness – that which was turned within as that which was turned without – lay dreaming or half awake beneath a common veil. The veil was woven of faith, illusion, and childish prepossession, through which the world and history were seen clad in strange hues. Man was conscious of himself only as a member of a race, people, party, family, or corporation – only through some general category. In Italy this veil first melted into air; an objective treatment and consideration of the State and of all the things of this world became possible. The subjective side at the same time asserted itself with corresponding emphasis; man became a spiritual individual, and recognized himself as such. In the same way the Greek had once distinguished himself from the barbarian, and the Arab had felt himself an individual at a time when other Asiatics knew themselves only as members of a race. It will not be difficult to show that this result was due above all to the political circumstances of Italy.[25]

Often taken as a summary of the Burckhardtian thesis, this passage certainly captures the pivotal relation of external state to internal character. But, as Burckhardt went on to emphasise, these elements of human consciousness have an ideal existence in a state of complementarity not competition. *L'uomo universale*, the all-sided or many-sided man, equalled the highest form of individual development as represented in the most powerful and varied nature; moreover, this multidimensional human being 'belonged to Italy alone'.[26] The repetition of the uniqueness of the Italian civic republican experience allowed Burckhardt to press home the most significant point of the book, namely 'the Renaissance would not have been the process of world-wide significance which it is, if its elements could so easily be separated from one another'; and he continued, 'it was not the revival of antiquity alone, but its union with the genius of the Italian people, which achieved the conquest of the Western world'.[27]

ii

Burckhardt's writing about the intellectual world of Renaissance Italy crafted a densely woven tapestry from the lives of individuals in the past and, as John Burrow has explained, this kind of approach relies on 'point[s] of intersection' that enable both detachment from and continuity with the present.

> Any rich and original work of historiography ... is a point of intersection: the simultaneous meeting place of the historian's unique personality, shaped by an intellectual and social milieu, with historiographical and literary models and conventions, with traditional and contemporary views of past events, and with a body of primary sources, some already established as important, others discovered or selected by the historian himself.[28]

To Burckhardt, the Renaissance constituted an original and extraordinarily complex moment in the history of Western civilisation, a moment that generated patterns, beliefs, and symbols, from its social customs, political institutions, and literary, artistic and architectural forms, which have become enmeshed in modern culture. So, like Carlyle before him, Burckhardt perceived history as an expression of the diversity of human experience, personified by the lives of great men and dramatised through their influence on political and social life. And, like his close contemporaries John Ruskin (1819–1900) and Walter Pater (1839–1894), he comprehended the writing of history as commenting on, as well as contributing to a vital discursive tradition, which involved explaining how a concept of the past is acquired, clarified and structured, and then suggesting how it is inflected in contemporary art and culture.

Axiomatic to the writings of Ruskin and Pater and Burckhardt was a universalistic thesis, which meant that they identified the truth of historical experience as embedded in a specific aspect of human activity, spiritual, natural, cultural or beautiful. All three writers seemed to believe that knowledge of the rise and fall of previous civilisations could provide instructive lessons for present social conditions, and especially the progress of culture. The relation of art, and especially beauty, to civilisation was central to this nineteenth-century narrative of improvement because it seemed to enable the translation of established modes of political representation into emerging forms of aesthetic understanding, especially with respect to the crucial notions of judgement, agency and instinct. The material worth and cultural renown of the beautiful was seen as necessarily contingent upon its historical framework and, therefore, ancient or medieval or Renaissance civilisations were invoked as a means to ameliorate the problems of rapidly industrialising society.[29] It should not surprise us, therefore, that while Burckhardt and Pater investigated the significance of the Renaissance as an historical phenomenon with far-reaching consequences for the development of art and culture, Ruskin referred to the Middle Ages in his desire to demonstrate the centrality of landscape art as a cultural text to the experience of the industrial world of nineteenth-century Britain.[30] Yet, regardless of whether the spotlight was placed on the growth of individuality (Burckhardt) or the cultivation of fidelity (Ruskin) or the expansion of sympathy (Pater), the aim of each writer was to anchor the present to the past in order to illuminate historical understanding as a form of human consciousness.

In *Modern Painters* (1843–60), Ruskin characterised the history of art in terms of a series of 'patriarchal' relationships, which had, he believed, the unfortunate result of presenting modern artists, like Turner and Landseer, as generally inferior to the old masters, such as Canaletto and Poussin.[31] Ruskin did not seem to think it was meaningful to posit strong claims about the continuity of thought and feeling between past and present histories of art; instead, he identified an essential dissimilarity in style and content from the great painting of past ages to modern art, and proposed that the success (and progress) of modern art depended upon its capacity to convey the eternal truths of nature and so delineate new standards of taste and judgement, especially with regard to landscape painting.[32] The 'Preface' to the first volume of *Modern Painters* (1843) makes explicit the urgent need for a new understanding of great art:

> [W]hen *public* taste seems plunging deeper and deeper into degradation day by day, and when the press universally exerts such power as it possesses to direct the feeling of the nation more completely to all that is theatrical, affected and false in art ... It becomes the imperative duty of all who have any perception or knowledge of what is really great in art, and any desire for its advancement in England, to come fearlessly forward ... to declare and demonstrate, wherever they exist, the essence and the authority of the Beautiful and the True.[33]

The only way to remedy this sorry state of affairs was, according to Ruskin, to redirect attention to the general principles and universal truths of the natural world as conveyed in painting, that is to say truths of tone, colour and chiaroscuro as well as of space, skies, clouds, mountains, water and vegetation. It was essential, he claimed, to compare the great works of ancient and modern landscape art and 'to show the real relations, whether favourable or otherwise, subsisting between it and our own'.[34] Not only the historical painters of the fifteenth century (Michelangelo and Leonardo da Vinci) but also, and perhaps more significantly, some modern landscape artists outranked the landscape painters of the seventeenth century (Claude and Gaspar) in their command of the technical principles of conception and composition. Ruskin's goal, therefore, was to demonstrate that greatness in art transcended historical periods and theoretical boundaries: 'the art is greatest which conveys to the mind of the spectator ... the greatest number of the greatest ideas'.[35]

Modern Painters proceeded far beyond its original intention – namely, a defence of J. M. W. Turner as a landscape painter of the highest order – to become a weighty treatise on the natural, beautiful and expressive in art. It attempted to endow the educated observer, rather than the connoisseurs and critics of art, with the special responsibility to legislate for the truth of art in relation to nature. Given that ideas of power, imitation, truth, beauty and relation could all be communicated by art, Ruskin posed a simple test for their veracity that required even the most inexperienced observer to ask whether what they saw captured in art, and in particular painting, could be corroborated in nature: 'Is it so? Is this the way a stone is shaped, the way a cloud is wreathed, the way a leaf is veined?'[36] The strength of this line of argument was that it provided an absolute aesthetic standard based on truth to nature and therefore bestowed a kind of cultural competence on those who chose to adopt this mode of looking. In *The Stones of Venice* (1851–53), for example, Ruskin outlined the principles of architecture and suggested a correlation could be drawn between the character of the builder and the political conditions of the Venetian state. Each stage in the building of St Mark's Cathedral in Venice embodied a physical characteristic and a moral quality so that discriminating the differences between its Byzantine and Gothic and Renaissance architecture required an understanding of the categorical shift from purism and naturalism to sensualism. To put it in other words, the Byzantine and the Gothic styles represented something like a prelapsarian state, which was followed by the fall of the Renaissance for, Ruskin argued, Venice had declined in power and prestige as a result of the Renaissance style of luxury and excess over goodness. The criteria upon which judgements of 'goodness' should to be made were defined in some 'notes relating to the competitive designs submitted for the new government buildings' (1853):

> To determine the goodness of any piece of art ... we have always to ask these two things: – First, does it do its practical work well: – and produce what we want , without loss of labour – if over a time: – Secondly, does it tell us what it is good for us to know: – and make us feel what it is good for us to feel.[37]

Like a plumb-line or spirit level used to measure alignment or balance, then, Ruskin's notion of a 'law of right' instituted a natural scale that could be universally applied to the form and content of paintings (as well as the shape and structure of buildings) in order to comprehend the moral value of things and so distinguish the good and the beautiful from the bad and the ugly.

For Pater, the authority of nature was of infinitely less importance than the impression of beauty that a picture, landscape, or personality produced, not as an abstract quality but a subjective response to the excellent work that could be found in all ages. In *Studies in the History of the Renaissance* (1873), he investigated the nature of our responses to all things beautiful, reflecting on the form and content of beauty as well as exploring the correspondence between beauty in art and nature.[38] Like Ruskin, Pater stressed the primacy of the perceiving subject in receiving the pleasures of sense but, unlike Ruskin, he suggested that these pleasures are the product of all the senses, taste, smell and touch as well as sight and hearing. The impression of beauty was crucial, Pater claimed, and so the aesthetic critic was charged with the task not of investigating the essential forms of beauty but of discerning the affective capacity of an individual subject to respond to beautiful things:

> What is important ... is not that the critic should possess a correct abstract definition of beauty for the intellect, but a certain type of temperament, the power of being deeply moved by the presence of beautiful objects. He will remember always that beauty exists in many forms. To him all periods, types, schools of taste, are in themselves equal. In all ages there have been some excellent workmen, and some excellent work done.[39]

This attempt to discriminate the physical sensibilities and mental impressions of human experience defined the aesthetic view of life advocated in *The Renaissance*, and it was crystallised in spectacular fashion in Renaissance Italy, of course. And yet, by characterising the Renaissance in the most capacious terms, from the end of the twelfth century in France to the eighteenth century in Germany, Pater offered a highly creative rendition of the life of the past, an imaginative synthesis of 'theories which bring into connexion with each other modes of thought and feeling, periods of taste, forms of art and poetry'.[40] Viewed in this way, as a series of moments or intervals to be expanded so that experience can be relished in all its dimensions, the Renaissance became less a general historical category and more a personal possession, expressing a renewed sensibility to art and culture.

A sense of the enlightening, and perhaps even emancipatory, capacity of the beautiful was absolutely crucial to Pater's intellectual project because he sought to establish individual subjective experience as the source of knowledge. But the pleasures of sense – sight, hearing, taste, smell and touch – were neither necessarily inferior nor inevitably corrupted according to Pater. Using the Greeks as the ideological base for his writing, and Greek as well as Renaissance art for its figurative examples, he advocated a philosophy of sensation that brought together the sensuous and the intellectual, organising feelings into a description of perception and taste and judgement. The consequence, he suggested, was that a language of inner sense referred to the structure of sensation as both appetite (desire) and emotion (feeling) in equal parts. One can think here of the 'Conclusion' to *The Renaissance*, which compared the outward with the inward life in order to develop a carefully wrought antinomianism directed towards the enriching of human experience:

> The service of philosophy, of speculative culture, towards the human spirit, is to rouse, to startle it to a life of constant and eager observation. Every moment some form grows perfect in hand or face; some tone on the hills or sea is choicer than the rest; some mood of passion or insight or intellectual excitement is irresistibly real and attractive to us, – for that moment only. Not the fruit of experience, but experience itself, is the end. A counted number of pulses only is given to us of a variegated, dramatic life.[41]

In other words, Pater held to the claim that the experience of beauty allowed for knowledge of the soul but he did not concede that the soul required advancement from a degraded, physical state. Rather, he dramatised beauty as a liberating force that enabled freedom through enjoyment but he also implied that this brought with it certain kinds of responsibilities; Pater's notion of beauty was a many-sided concept, individual and variegated and transient, which enabled the freedom of the individual to judge subjects and objects (including artworks) for themselves apart from the (distracting) pressures of the world.

iii

The discussions on art and culture by Burckhardt, Ruskin and Pater express a related interest in the question of whether beauty should be considered as a relative phenomenon, embedded within a specific historical context as one of many parts of the complex web of the past, or an exemplary concept, detached from any sort of context and instead performing something like the role of an independent guide to systems of thought and feeling. So, whereas Burckhardt isolated the emergence of individualism as a 'new fact ... in history' that fostered a vital link between art (and in particular beauty) and the state; Ruskin identified the forms of nature as the essential sourcebook for the discovery of beauty and therefore truth about culture and society; and Pater stressed the importance of a multiplied consciousness of self as the catalyst for appreciating 'a philosophy of the variations of the beautiful' in art as in life.[42] All three writers advanced teleological narratives of the progress of culture towards a high point of human consciousness – the Italian Renaissance for Burckhardt, medieval Gothic for Ruskin, and Ancient Greece for Pater – and included the selection of great men to illustrate their narratives with examples of the universal power of genius; namely Dante, Turner and Plato, respectively. Yet it was Burckhardt who offered the clearest vision of the future prospects for the study of art in an unpublished manuscript entitled, 'Introduction to a universal art history' (c. 1887–90). On the one hand, he wrote, art history 'sets out essentially to be the handmaiden of history, and of cultural history in particular, and thus to regard beauty as an element inseparable from each specific historical context'; and on the other, 'it takes beauty itself as its starting point, and enlists cultural history only as an aid. Both have their justifications'. Pragmatically, therefore, the solution was simple: 'we shall have to take a middle way'.[43]

Following Burckhardt, and quite deliberately so, Aby Warburg (1866–1929) mapped out a synthetic 'middle way' for the history of art, balancing the existing tendency to concentrate on the stylistic features of painting with an emerging interest in depicting the intellectual and social milieux of artists and their works.[44] Warburg translated his own fascination with Renaissance art, and in particular the ways in which the Florentines of Lorenzo de Medici's circle identified with and were influenced by the forms of antiquity, into a more general mode of enquiry that attempted to illuminate the processes through which symbols fundamentally associated with a pagan world were translated into the religious culture of fifteenth-century Italy. As a result, his project was double-edged: to establish a programme for studying the memory of the past as it impacts upon the formation of culture, and to illustrate the influence of antiquity in all its dimensions (social, political, religious, scientific, philosophical, literary and artistic) on modern European civilisation. Delivering a keynote paper to the tenth Art-Historical Congress in Rome (October 1912), Warburg proposed an extension of the borders erected around the new discipline as a means of reviewing and revising the kind of intellectual work deemed to fall within the scope of art-historical enquiry. He explained:

> Until now, a lack of adequate general evolutionary categories has impeded art history in placing its materials at the disposal of the – still unwritten – 'historical psychology of human expression' ... By attempting to elucidate the frescoes in the Palazzo Schifanoia in Ferrera, I hope to have shown how an iconological analysis that can range freely, with no fear of border guards, and can treat the ancient, medieval, and modern worlds as a coherent historical unity – an analysis that can scrutinise the purest and most utilitarian of arts as equivalent documents of expression – how such a method, by taking pains to illuminate one single obscurity, can cast light on great and universal evolutionary processes in all their interconnectedness.[45]

Most of the aspects of the work undertaken by Burckhardt, as well as Ruskin and Pater, are here made explicit, including the free movement of ideas between different areas of intellectual enquiry, the striking continuities of thought between distinct historical periods, and the subtle correspondences between discrete cultural formations. For, as the passage indicates, Warburg was centrally concerned with the hermeneutic problem of how to make sense of art works, not only the proper materials for their analysis but also the barriers to achieving a proper cultural and historical interpretation of them.

Rejecting the evolution of art as a straightforwardly progressive movement, Warburg advocated a theory of artistic development that included dramatic breaks in the continuum of history against prevailing trends, as well as the repetition of themes emphasising specific trends. Thus he asked a difficult question: 'to what extent can the stylistic shift in the presentation of human beings in Italian art be regarded as part of an international process of dialectical engagement with the surviving imagery of Eastern Mediterranean pagan culture?'[46] A new way of studying art, described as the 'historical psychology of human expression', would provide the means of answering this question by integrating the two main interpretive orthodoxies about art, as a contextualised field or an autonomous sphere, and so formulating a distinctive method for explicating artworks. Thus, works of art would be understood as both *reflecting* wider cultural and political discourses and *determining* the scope of their own representational agenda. So, far from accepting a view of art detached from culture and society, Warburg's intervention into discussions about the form and content of art-historical study depended on a vision of art as participant in as well as indicator of the reflective and expressive life of the past. In sum, he mapped out the route for a distinctive programme of art-historical study – truly the 'middle way' for the history of art imagined by Burckhardt – which involved the disciplines of history and psychology, and emphasised 'coherence', 'equivalence' and 'interconnectedness'.

The legacy of Warburg's theory of art was felt on many different levels and across many scholarly fields of enquiry, especially in the methodologies that have been developed to make sense of artworks as cultural products and social and political texts. The Warburgian approach proved crucial not only to the development of art history as a hermeneutical discipline but also to reaffirming the value of humanistic enquiry within and across different disciplines. The establishment of the Warburg Institute, 'to further the study of the classical tradition, that is of those elements of European thought, literature, art and institutions which derive from the ancient world', developed directly out of Warburg's research ideas and method. His personal library provided the intellectual foundations for a research institute in cultural history, established in Hamburg in 1921 under the directorship of Fritx Saxl (1890–1948) and then moved to London in 1933 in order to escape the Nazi regime; and his work on the history of art inaugurated a tradition of scholarship that sought to understand the 'interconnectedness' of the history of Western civilisation.[47] The Warburg library literally embodies the kind of work now called cultural history,[48] with its four organising categories of Action, Orientation, Word and Image, and its ambitious mission to study the survival and transformation of ancient patterns in social customs and political institutions; the gradual transition, in Western thought, from magical beliefs to religion, science and philosophy; the persistence of motifs and forms in Western languages and literatures; and the tenacity of symbols and images in European art and architecture.[49]

Yet London presented a very different kind of intellectual setting for the Institute compared to Hamburg; there was considerable division between academic disciplines (and art history was a relatively new discipline) and, perhaps more importantly, the Institute did not find a permanent home until its move to Woburn Square in 1958. In other words, there was an urgent need for stability and coherence if the aim of demonstrating and disseminating a cross-disciplinary approach to the study of history, and writings about history, was to be achieved.

Two aspects of the early life of the Warburg Institute in London were crucial in establishing firm foundations for the kind of history of art that Warburg had proposed. First, the creation of the *Journal of the Warburg Institute*, pioneered by Saxl and his colleagues, in order to establish the intellectual foundations as well as to promote the values of, and stimulate debate about, European humanism against the rise of National Socialism in Germany and Italy. With the ambition 'to take the study of Humanism, in the broadest possible sense, for its province', the founding editors, Edgar Wind (1900–1971) and Rudolf Wittkower (1901–1971), outlined the ideological agenda of the *Journal* as follows: it would 'supply a common forum for historians of art, religion, science, literature, social and political life, as well as for philosophers and anthropologists', and 'explore the workings of symbols – the signs and images created by ancient, and employed by modern generations, as instruments both of enlightenment and superstition'.[50] Second, and linked directly to the *Journal*, the association of the Warburg with the Courtauld Institute of Art developed out of the shared sense amongst members of the profound importance of the classical tradition, and especially its formative influence upon Renaissance art and culture. The connection with the Courtauld Institute, established in 1932 (by Viscount Lee of Fareham, Samuel Courtauld and Sir Robert Witt) in order to improve knowledge and understanding of the visual arts in Britain,[51] was formalised in the third volume in 1940 when the title was changed to the *Journal of the Warburg and Courtauld Institutes* in order to acknowledge the intimate connections between the two Institutes.[52]

The affiliation of the Warburg and Courtauld Institutes was immensely significant for the humanistic study of the arts because it went against the grain of the increasing specialisation and related separation of academic disciplines and demonstrated, instead, the value of an institutional space where the 'interconnectedness' between (cultural and intellectual) history and (the history of) art was the primary mode and object of study. Take, for example, the work of Ernst Gombrich (1909–2001), which exemplified the scope of this cross-disciplinary mode of scholarly enquiry. Director of the Warburg Institute from 1959 to 1976, Gombrich's writings on visual perception and pictorial representation conveyed the main characteristics of the shared intellectual project of the Institutes, namely a commitment to the historical framing of ideas about art and culture, a willingness to roam across disciplinary borders, and a desire to communicate in a language accessible to a wide, and not necessarily specialist, readership. Of course, other members of the Institutes were also influential in forging links between intellectual history and the history of art, including such luminaries as Francis Wormald, Francis Yates, T. S. R. Boase and Anthony Blunt. But it was Gombrich, above all, who exerted the strongest influence on the future direction of interdisciplinary study through his determination to use ideas about perception, derived from psychology and philosophy, in order to investigate the very nature of representation in art. What happens when we look at a painting? In what ways does the medium of painting influence our interpretation of its form and content? And, equally, to what extent does the painter simply depict what they see in nature?

Gombrich's research was driven by a belief that psychological factors mattered in shaping the conventions of the history of art. *The Story of Art* (1950) was his first, and still hugely influential, study of image-making in which he drew on the long-established tradition in art history of distinguishing between 'knowing' and 'seeing', or what can be known about the visible world compared to what can be seen in it. Gombrich sought to identify the conventions that have surrounded and helped define artworks: it is precisely because a painter's style becomes freighted with historical conventions and attributions in the name of truthfulness to a specific vision of things that, he claimed, 'different ages and different nations have represented the visible world in such different ways'.[53] Using a teleological narrative in order to make his points, he considered the evolution of art from the primitives and Egyptians to the Impressionists in terms of a shift in modes of representation from 'what they knew' to 'what they saw'.[54] Primitive artists emphasised form over content and so they did not copy actual faces but drew shapes. In a similar vein, the Egyptians represented what they knew about, rather than saw in, the world; and, subsequently, Greek and Roman art followed by medieval and Chinese art turned these conceptual forms into representations of everyday life, sacred story, and contemplation respectively. According to Gombrich, the recommendation that artists should paint what they see was an invention of the Renaissance, correlating to advances in the science of perspective together with a new understanding of colour, movement and expression. But, as he showed, this emerging notion of perception pushed against established artistic conventions until a point in the late nineteenth century when the Impressionists proclaimed their paintings were free from convention and, instead, conveyed what they really saw with the accuracy of science. Gombrich contended that artists could not record what they saw without referring to conventions of style, and so, regardless of the appeal of Impressionist art, the claim upon which it was founded was only partially true – artists cannot work in isolation from their cultural conditions, literally painting what they see, because what is seen in the world cannot be split from what is known about it.

Having suggested the weakness of an art founded on seeing rather than knowing, Gombrich turned his attention to theories of perception and the changing styles of representation. He delivered the A. W. Mellon Lectures in the Fine Arts at the National Gallery in Washington on the subject of 'The Visible World and the Language of Art' (1956), and later expanded these speculations into a book-length study that revised his earlier story of visual discoveries. *Art and Illusion: A Study in the Psychology of Pictorial Representation* (1960) examined the history of representation as illuminated by the different styles of art and pondered the difficulty of comprehending 'why it should have taken mankind so long to arrive at a plausible rendering of visual effects that create the illusion of life-likeness'.[55] Revisiting (in order to amend) the ideas about 'seeing' and 'knowing' advanced in *The Story of Art*, Gombrich argued that painters are translators rather than transcribers of the world, balancing what they know about the world with what they see in a complex process of 'making and matching'.[56]

> The limits of likeness imposed by the medium and schema, the links in image-making between form and function, most of all, the analysis of the beholder's share in the resolution of ambiguities will alone make plausible the bald statement that art has a history because the illusions of art are not the only fruit but the indispensable tools for the artist's analysis of appearances.[57]

Citing John Ruskin and Roger Fry as apologists for the 'truth to nature' version of painting, Gombrich explained that the central problem in the history of art is that 'our knowledge of the visible world' is contrapuntal to the 'rendering of a three-dimensional world on a flat canvas'. In other words, painting might appear to be reliant on 'forgetting what we know about the world' but it actually rests on the knowing illusion of 'inventing comparisons that work'.[58]

iv

Theodor Adorno described the practice of the Warburg Institute as concerned with 'the afterlife of classical antiquity', a phrase that astutely pinpoints the intellectual project of humanism with its emphasis on the continuous process of self-knowledge and self-realisation.[59] The notion that we retreat into the past as refuge from the ills of the present, clinging onto the vestiges of earlier, and putatively less discontented, civilisations, does not fit with the ambitions of the Warburg Institute. Those ambitions, shaped by Burckhardt and Warburg and sustained by Gombrich, were to offer a humanistic education that explores how history is made and shaped in and through the 'interconnectedness' of ideas about art and culture. Cultural history is the name most often used to define this interconnected (or cross-disciplinary) approach to the thought and feeling of the past: it is located in the interstices of intellectual history and the history of art, and so involves an undertaking to chart the ideas, patterns, motifs and images that have survived in, and transformed, Western intellectual traditions. A cultural historian teaches us to isolate and identify the concepts and themes articulated in historical writings on ideas and images, seeking to understand the cultural formations produced at the intersections of distinct areas of intellectual enquiry while also remaining attentive to the nuances and subtleties of different discursive genres.

In a significant sense, Burckhardt, Ruskin and Pater should be considered cultural historians, anticipating the Warburgian tradition in their view of the life of the past as imbricated with the experience of the present, and their perception of artworks as expressive of culture and society. A direct line of thought can be traced from Burckhardt to Gombrich via Warburg, and a parallel, but no less important, line can be mapped from Ruskin to Gombrich via Pater. These writers have a profound impact on the establishment of the history of art and intellectual history, and the subsequent development of cultural history, as the family relations of an intellectual tradition of historical writing that is held together by a commitment to humanism. However, humanism has come under sustained attack in the last 30 or so years as a result of the rise of (structuralist and post-structuralist) critical theory – and subsequently post-colonial studies – and the powerful impact of feminism. The problem is that although humanism has sought to explore everything that is human, and explain what makes the human special, its frame of reference is usually delineated such that it focuses on elite forms of knowledge and high types of culture and, moreover, expresses the views of a masculine establishment. I want to conclude, therefore, by considering the role of humanism, and the humanities, in the twenty-first century, looking especially at three influential examples of interdisciplinary work that is inflected to greater and lesser degrees by the history of art and intellectual history, and therefore demonstrates the scope and limitations of humanistic enquiry.

For Michael Baxandall, the history of art is fundamentally a social history; that is to say, it is concerned with the practices and procedures encountered and enunciated in the everyday life of a given society: 'the *style* of pictures is a proper material of social history. Social facts lead to the development of distinctive visual skills and habits: and these visual skills and habits become identifiable elements in the painter's style.'[60] In *Painting and Experience in Fifteenth-Century Italy* (1972), for instance, he shows 'what it was like, intellectually and sensibly, to be a Quattrocento person'.[61] Baxandall's early work is motivated by a desire to understand in what ways, and to what extent, the visual skills and habits that develop in a society are reflected in a painter's style. Very much in the Warburgian tradition, then, he concentrates on the Renaissance (like Burckhardt) and examines theories of perception (like Gombrich), but he also introduces new ideas about what, and who, constitutes 'the social' into the idiom of this established tradition. A vernacular mode of using visual conventions is embedded in the social practice of the Quattrocento, he claims, and so historical records of commercial transactions and scientific knowledge about vision and images are claimed as vital sources in explicating the forms and styles expressed in a fifteenth-century painting. The relationship between painting and social, religious and commercial life is not that of one-way traffic, however; painting is fashioned from these social relations and yet, at the same time, it offers a cogent critique of them. This sense of art as reflecting and determining its cultural conditions underpins Baxandall's subsequent work on representation; consequently, *Patterns of Intention* (1985) analyses the painter's role in the formation of visual culture, while *Shadows and Enlightenment* (1995) examines the correspondence between ideas about shadows and our visual experience of the world.[62]

A similar position on painting is articulated by John Barrell in his studies of eighteenth-century culture, and in particular *The Dark Side of the Landscape* (1980) and *The Political Theory of Painting, from Reynolds to Hazlitt* (1986).[63] Barrell, too, is preoccupied with the capacity of painting to express the experience of everyday life – and, consequentially, the potential for disjunction between artistic conventions and social reality – and he also perceives art as central to our understanding of the past. However, Barrell interprets the notion of an image in its widest sense, taking it to mean literary, political, social and moral forms as well as visual ones, or, in other words, as a multicultural mode of representation rather than simply a painterly one. Inspired by E. P. Thompson's *The Making of the English Working Class* (1963), Barrell discusses the apparent stability of the image of rural life in the paintings of Gainsborough, Morland and Constable in *The Dark Side of the Landscape*:

> The painting [of rural life], then, offers us a mythical unity and – in its increasing concern to present an apparently more and more actualised image of rural life – attempts to pass itself off as an image of the actual unity of an English countryside innocent of division. But by examining the process by which that illusion is achieved – by studying the imagery of the paintings, the constraints upon it, and upon its organisation in the picture-space – we may come to see that unity as artifice, as something made out of the actuality of division.[64]

Unlike Baxandall, Barrell emphasises the ideological systems of thought at work in a specific culture and society as a means of elucidating the politics of the division of labour. His point is that we are just as implicated now (as disinterested but admiring observers) in perpetuating the illusion of unity and harmony in labour as the connoisseurs of the eighteenth century were then (as patrons and interested observers). Barrell's subsequent work develops these themes, analysing the idea of 'the public' as a political term indexed to citizenship. In *The Political Theory of Painting*, he argues that the discourse of (what is now called) civic humanism conceptualised polite society in profoundly political terms as a republican state: it sought to cultivate the public virtues of the republican citizen over and above their private pleasures and, crucially, painting played a leading role in this cultivating process.[65] The function of painting in commercial society was, ostensibly, that of 'describing and recommending those virtues which will preserve a civil state, a public, from corruption'.[66] But, contends Barrell, the attempt to bestow a political purpose upon painting as the expression of a 'grand style', principally by Sir Joshua Reynolds, was abandoned by Hazlitt, who asserted 'that the ends and satisfactions of painting were primarily private' while also rejecting 'the traditional interdependence of the republic of taste and the political republic'.[67]

Barrell and Baxandall offer compelling narratives of given historical periods (the fifteenth and eighteenth centuries, respectively) that derive their coherence from a perception of art as one kind of cultural formation. Much of their work is an attempt, therefore, to formulate explanations of what can be learnt from images (visual or otherwise) about the lived experience of the past. Yet, and significantly, by sketching the contours of an elite culture and, at the same time, expounding the tenets of a masculine aesthetic, Baxandall and Barrell discuss noticeably privileged versions of the development of individuality, and principally as articulated by white, male, bourgeois subjects.[68] Thus my last example is that of Ludmilla Jordanova, whose work on science and sexuality has done much to amplify the often one-sided orthodoxies advanced in the name of humanistic enquiry and promote, instead, a many-sided interpretation of human nature. Setting out the rationale for *Sexual Visions* (1989), Jordanova likens the task of an historian to the opening of a box containing a jumbled assortment of things that together define its purpose; the historian of science and medicine must 'unpack the processes through which "naturalization" takes place, whereby ideas, theories, experiences, languages, and so on, take on the quality of "being natural"'.[69] Highlighting the images, myths and metaphors about gender utilised by a range of authors over more than 200 years, she proposes that science and medicine act as systems of representation within a cultural domain. Far from removing the boundaries that separate science and medicine from other areas of intellectual enquiry, though, Jordanova maintains that interdisciplinary research relies on the sort of cultural analysis that can adumbrate the distinctiveness of certain privileged kinds of knowledge as a means towards recognising the mystifications they produce.

What we learn from Jordanova's work is that intellectual care and scholarly sophistication are needed to move between different types of knowledge (privileged and popular), and distinct cultural domains (high and low). Recently, she has redefined the scope of her work, incorporating the focus on images and representations into the more general, and perhaps less generic, descriptive category of visual culture, which 'implies, for example, both that the variety of cultural forms around at a given moment should be considered and that traditional assessment of aesthetic quality need not be the primary consideration when selecting sources to study'.[70] It is a significant shift because it shows how far we have come along the route of 'interconnectedness', from Burckhardt, Pater and Ruskin, and the activities of the Warburg Institute, in order to arrive at current academic work that challenges the established goals of a humanistic education to expand and include considerations of society, class and gender. Humanism is not inevitably incompatible with the demands of life in the twenty-first century; nonetheless, as Edward Said so eloquently explains, it is the public duty of intellectuals, as citizens, to revise its language and representations to meet the demands of a democratic world:

> Humanism is not about withdrawal and exclusion. Quite the reverse: its purpose is to make more things available to critical scrutiny as the product of human labour, human energies for emancipation and enlightenment, and just as importantly, human misreadings and misinterpretations of the collective past and present. There was never a misinterpretation that could not be revised, improved, or overturned. There was never a history that could not to some degree be recovered and compassionately understood in all its suffering and accomplishment. Conversely, there was never a shameful secret injustice or cruel collective punishment or a manifestly imperial plan for domination that could not be exposed, explained, and criticized.[71]

Implicit in Said's new manifesto for humanism is something very like the palimpsest metaphor, with which this chapter began, as a means of understanding the layer upon layer of writing that makes up our collective memory. More than any other metaphor for historical understanding – intellectual, artistic or cultural – the idea of history as palimpsest conveys the profound but necessary labour involved in interpreting and explaining the life of the past in the present.

notes

1. Thomas Carlyle, 'On History', *Fraser's Magazine*, No. 10 (1830); reprinted in *Critical and Miscellaneous Essays* (1839); *The Works of Thomas Carlyle*, 'The Ashburton Edition' (17 Vols, London, 1885–88), XV, p. 499. Carlyle returns to the subject in an inaugural lecture for the *Society for the Diffusion of Common Honesty*, advocating the profit to be acquired from the study of history if, and only if, the past can be compressed so that it represents 'Universal History' as a 'magic web'; a fragment of this lecture was published as 'On History Again' in *Fraser's Magazine*, No. 41 (1833); rpr in *Works*, XVI.

2. Carlyle, 'On History', p. 499.

3. See, in particular, Chris R. Vanden Bossche, *Carlyle and the Search for Authority* (Columbus, Ohio, 1991).

4. The classic work on the methodology of history is Herbert Butterfield's *The Whig Interpretation of History* (London, 1931), which offers a persuasive critique of discipline history. Important contributions to this discussion are also made in R. G. Collingwood's *The Idea of History* (Oxford, 1946), and E. H. Carr's *What is History?* (London, 1961).

5. More recent contributions to the debate about the scope and limitations of historical study include: J. Appleby, Lynn Hunt and M. Jacob, *Telling the Truth about History* (New York, 1994); P. Burke, ed., *New Perspectives on Historical Writing* (Cambridge, 1991); Lynn Hunt, ed., *The New Cultural History* (Berkeley, Calif., 1989); M. C. Lemon, *The Discipline of History and the History of Thought* (London, 1995), and John Tosh, *The Pursuit of History: aims, methods and new directions in the study of modern history* (London, 1984).

6. On the regulatory practices of public institutions and the disciplining of society, see Michel Foucault, *Discipline and Punish: the birth of the prison*, trans. Alan Sheridan (1975; Harmondsworth, 1991).

7. See Amanda Anderson and Joseph Valente, eds, *Disciplinarity at the Fin de Siècle* (Princeton, NJ, 2002). The introduction is especially helpful in outlining the closed compared to the open model of disciplinary formation.

8. A useful guide to historiography is Michael Bentley, ed., *Companion to Historiography* (London, 1997).

9. For those interested in pursuing these matters, the following suggestions should prove helpful: Svetlana Alpers, *The Art of Describing* (Chicago, Ill, 1983); Thomas Crow, *Emulation: making artists for revolutionary France* (New Haven, Conn., 1995); Michael Fried, *Courbet's Realism* (Chicago, Ill, 1990); Francis Haskell, *History and its Images: art and the interpretation of the past* (New Haven, Conn., 1993); Marcia Pointon, *History of Art: a student's handbook* (London, 1980), and Alex Potts, *Flesh and the Ideal: Winckelmann and the origins of art history* (New Haven, Conn., 1994).

10. See, for example, John Burrow, *A Liberal Descent: Victorian historians and the English past* (Cambridge, 1981); Donald Winch, *Riches and Poverty: an intellectual history of political economy in Britain, 1750–1834* (Cambridge, 1996), and Stefan Collini, Richard Whatmore and Brian Young, eds, *History, Religion, and Culture: British intellectual history, 1750–1950* (Cambridge, 2000).

11. Stefan Collini, Donald Winch and John Burrow, *That Noble Science of Politics: a study in nineteenth-century intellectual history* (Cambridge, 1983), pp. 5–6.

12. The phrase is Stefan Collini's, from his important and original study of the interrelationship of political debates, social attitudes and aesthetic judgements in Victorian society: *Public Moralists: political thought and intellectual life in Britain, 1850–1930* (Oxford, 1991), p. 1.

13. I have found Alan S. Kahan's recent book, *Aristocratic Liberalism: the social and political thought of Jacob Burckhardt, John Stuart Mill, and Alexis de Tocqueville* (New Brunswick, NJ, 2001), especially helpful in thinking about the ideological underpinnings of Burckhardt's historical project. Also worth consulting are: John Roderick Hinde, *Jacob Burckhardt and the Crisis of Modernity* (Montreal, 2000), and Richard Sigurdson, *Jacob Burckhardt's Social and Political Thought* (Toronto, 2003).

14. The publication history of Burckhardt's works is quite complex because, although no new work was published after 1867, a number of volumes of lectures and essays were published together with a study of the cultural history of Greece. In his

lifetime, Burckhardt produced *The Cicerone: a guide for the enjoyment of works of art in Italy* (1855; trans. 1873) and *History of Early Modern Architecture: the Renaissance in Italy* (1867), originally published as the first part of volume four of Franz Kugler's *History of Architecture* and then released as an independent volume, *History of the Renaissance in Italy* (1878). After his death in 1897, Hans Trog published a small volume, *Recollections of Rubens* (1898), and a compilation of essays, *Contributions to the History of Art in Italy* (1898); then Jacob Oeri, Burckhardt's nephew, published four volumes of Burckhardt's lectures, entitled *History of Greek Culture* (1898–1902), followed by a selection of his lectures on historical method, *Considerations on World History* (1905).

15. Roughly 15,000 copies were sold from 1878 to around 1900, and approximately 500,000 copies have been sold since 1928 when the copyright expired. I shall be referring to the Middlemore translation, reproduced in complete and unabridged form by L. Goldscheider (London, 1945).

16. Describing his own work, John Burrow offers a nice formulation of this principle, for, he explains, 'one of the ways in which a society reveals itself, and its assumptions and beliefs about its own character and destiny, is by its attitudes to and uses of its past' (*A Liberal Descent*, pp. 1–2).

17. Burckhardt, *Civilization of the Renaissance*, p. 1.

18. Ibid.

19. Ibid., p. 2.

20. Ibid., p. 44.

21. Ibid., p. 50.

22. Ibid., p. 43.

23. Ibid., p. 52.

24. Ibid., p. 57.

25. Ibid., p. 87.

26. Ibid., p. 90.

27. Ibid., p. 111.

28. Burrow, *A Liberal Descent*, p. 4.

29. Of course, the strongest influence on the nineteenth century came from Ancient Greece and the waxing and waning of interest in the subsequent medieval and Renaissance periods was directly indexed to the degree of reverence for the Greeks. The two classic studies of this influence are: Richard Jenkyns, *The Victorians and Ancient Greece* (Oxford, 1980), and Frank M. Turner, *The Greek Heritage in Victorian Britain* (New Haven, Conn., 1981).

30. There is a tendency in critical work to consider Ruskin and Pater as representatives of diametrically opposed views on art and culture; however, the following studies offer more nuanced interpretations of the debates about art in the period: Richard Dellamora, *Masculine Desire: the sexual politics of Victorian aestheticism* (Chapel Hill, NC, 1990); Linda Dowling, *The Vulgarization of Art: the Victorians and aesthetic democracy* (Charlottesville, Va, 1996); Kathy A. Psomaides, *Beauty's Body: femininity and representation in British aestheticism* (Stanford, Calif., 1997), and Jonah Siegel, *Desire and Excess: the nineteenth-century culture of art* (Princeton, NJ, 2000).

31. The context for Ruskin's approach to the history of art is explored in J. B. Bullen, 'Ruskin and the Tradition of Renaissance Historiography' in Michael Wheeler and Nigel Whitely, eds, *The Lamp of Memory: Ruskin, Tradition, and Architecture* (Manchester, 1992).

32. For some of the best critical studies on Ruskin, see Dinah Birch, *Ruskin's Myths* (Oxford, 1988); Elisabeth K. Helsinger, *Ruskin and the Art of the Beholder* (Cambridge,

Mass., 1982); Robert Hewison, *John Ruskin: the argument of the eye* (London, 1976), and George Landow, *The Aesthetic and Critical Theories of John Ruskin* (Princeton, NJ, 1971).

33. John Ruskin, *Modern Painters* (5 Vols, 1843–60), *The Collected Works of John Ruskin*, eds, E. T. Cook and Alexander Wedderburn (39 Vols, London, 1903–12), III, p. iv, 'Preface to the First Edition'.

34. Ruskin, *Collected Works*, III, p. 83. The full sentence is as follows: 'It is my purpose, therefore, believing that there are certain points of superiority in modern artists ... which have not yet been fully understood ... to institute a close comparison between the great works of ancient and modern landscape art; to raise, as far as possible, the deceptive veil of imaginary light through which we are accustomed to gaze upon the patriarchal work; and to show the real relations, whether favourable or otherwise, subsisting between it and our own.'

35. Ibid., p. 92.

36. Ibid., p. 180.

37. John Ruskin, 'Fragment of Notes Relating to the Competitive Designs Submitted for the New Government Buildings' (1853), *Huntington Library*, manuscript no. HM 21661.

38. The complexity of Pater's understanding of beauty, and its influence on debates about aestheticism, is nicely conveyed in the following studies: David J. DeLaura, *Hebrew and Hellene in Victorian England: Newman, Arnold, and Pater* (Austin, TX, 1969); Jonathan Freedman, *Professions of Taste: Henry James, British aestheticism, and commodity culture* (Stanford, Calif., 1990), and Carolyn Williams, *Transfigured World: Walter Pater's aesthetic historicism* (Ithaca, NY, 1989).

39. Walter Pater, *The Renaissance: studies in art and poetry. The 1893 text*, ed. Donald L. Hill (Berkeley, Calif., 1980), p. xxi. Four editions of *The Renaissance* were published in Pater's lifetime (1873, 1877, 1888, 1893), and a fifth edition was produced after his death as part of an eight-volume edition of *The Works of Walter Pater* (London, 1900–01); however, this set was superceded by a ten-volume Library Edition of *The Works of Walter Pater* (London, 1910), which has become the standard edition in use today. Donald Hill's new scholarly edition follows the text of Pater's own last revised version of *The Renaissance*, 'the one which thus embodies his own maturest judgement in matters of form, style, and meaning' (p. x); and includes textual and explanatory notes, together with translations, and a couple of Pater's book reviews from 1872. Importantly, the infamous 'Conclusion' first appeared as part of Pater's review of 'Poems by William Morris' in the *Westminster Review* (October 1868), but Pater removed it from the second edition of *The Renaissance* (1877) – seemingly because 'it might possibly mislead some of those young men into whose hands it might fall' – before reinstating it, with slight revisions, in the third (1888) and fourth editions (1893). See Hill, pp. 443–51, for a full account of this rather curious sequence of events.

40. Pater, *The Renaissance*, p. 2.

41. Ibid., p. 188.

42. Ibid., p. 103.

43. Cited by Maurizio Ghelardi in the introduction to a newly available translation of Burckhardt's manuscript on *Italian Renaissance Painting according to Genres*, trans. David Britt and Caroline Beamish (Los Angeles: The Getty Research Institute, 2005).

44. See E. H. Gombrich, *Aby Warburg: an intellectual biography* (London, 1970).

45. Aby Warburg, *The Renewal of Pagan Antiquity: contributions to the cultural history of the European Renaissance*, introd. K. W. Forster, trans. David Britt (Los Angeles, Calif., 1999), pp. 585–6. In an anniversary lecture, E. H. Gombrich examined Warburg's vision of 'the purpose of art history and the methods it should employ' and evaluated its lasting impact on the discipline: 'Aby Warburg: His Aims and Methods', *Journal of the Warburg and Courtauld Insitutes*, 62 (1999), 268–82.

46. Warburg, *Renewal of Pagan Antiquity*, p. 585.

47. A useful description and history of the Warburg Institute is given on their website at the following address: <http://www.sas.ac.uk/warburg/institute/institute_introduction.htm>.

48. Warburg's term was '*Kulturwissenschaft*', the notion around which his Hamburg library was organised and from which the project of the Warburg Institute gained its momentum. The Warburg Institute was always committed to interdisciplinary work but it was not until the early 1970s that the term 'cultural history' became a key part of the *Journal*'s mission statement.

49. See <http://www.sas.ac.uk/warburg/mnemeosyne/SUBJECTS.htm>.

50. See <http://www.sas.ac.uk/warburg/journal/historyjwci.htm>.

51. See 'A Short History of the Courtauld Institite of Art' by Peter Kidson at <http://www.courtauld.ac.uk/history.html>.

52. In its 65-year publication history, three volumes of the *Journal of the Warburg and Courtauld Insitutes* have acquired special significance: number 9 (1946) marks the immediate aftermath of the Second World War and so reaffirms the shared interests of British and European scholarship with a volume composed exclusively of contributions from Italy; number 35 (1972) signals a subtle recalibration of the aims of the *Journal* by its editors, David Chambers and J. B. Trapp, according to the principles of cultural history and an interdisciplinarity approach; and number 50 (1987) represents the fiftieth volume and so celebrates its enduring tradition of reconstructing conversations with the past via contributions from past and present members of the Institutes.

53. E. H. Gombrich, *The Story of Art* (London, 1950).

54. Ibid., p. 3.

55. E. H. Gombrich, *Art and Illusion: a study in the psychology of pictorial representation* (London, 1960), p. 157.

56. Ibid., p. 24. Outlining the purpose of the book, Gombrich explains: 'my main concern was with the analysis of image-making – the way, that is, in which artists discovered some of these secrets of vision by "making and matching" ... As a secular experiment in the theory of perception, illusionist art perhaps deserves attention even in a period which has discarded it for other modes of expression.'

57. Ibid., pp. 24–5.

58. Ibid., p. 254.

59. Theodor Adorno, *Aesthetic Theory*, ed. Gretel Adorno and Rolf Tiedemann, trans. Robert Hullot-Kentor (1970; London, 2002), p. 5. The full sentence is as follows: 'Tracing aesthetic forms back to contents, such as the Warburg Institute undertook to do by following the afterlife of classical antiquity, deserves to be taken more broadly.'

60. Michael Baxandall, *Painting and Experience in Fifteenth-Century Italy: a primer in the social history of pictorial style* (Oxford, 1972), 'Preface to the First Edition'.

61. Ibid., p. 152.

62. Michael Baxandall, *Patterns of Intention: on the historical explanation of pictures* (New Haven, Conn., 1985) and *Shadows and Enlightenment* (New Haven, Conn., 1995).

A special edition of *Art History*, 21:4 (1998), later published as a separate volume, is devoted to the substantial influence of Baxandall's work on a new kind of art history; see *About Michael Baxandall*, ed. Adrian Rifkin (Oxford, 1999).

63. John Barrell, *The Dark Side of the Landscape: the rural poor in English painting, 1730–1840* (Cambridge, 1980), and *The Political Theory of Painting, from Reynolds to Hazlitt – 'The body of the public'* (New Haven, Conn., 1986).
64. Barrell, *The Dark Side of the Landscape*, p. 5.
65. J. G. A. Pocock offers an immensely important account of the principles and practices of civic humanism in *The Machiavellian Moment: Florentine Republican Thought and the Atlantic Tradition* (Princeton, NJ, 1975).
66. Barrell, *The Political Theory of Painting*, p. 10.
67. Ibid., p. 64.
68. See Barrell's essay on '"The Dangerous Goddess": masculinity, prestige, and the aesthetic in early eighteenth-century Britain', *Cultural Critique*, 12 (1989), 101–31, for an insightful analysis of the complex relationship between these issues.
69. Ludmilla Jordanova, *Sexual Visions: images of gender in science and medicine between the eighteenth and twentieth centuries* (Hemel Hempstead, 1989), p. 5.
70. Ludmilla Jordanova, *History in Practice* (London, 2000), p. 89.
71. Edward Said, *Humanism and Democratic Criticism* (New York, 2004), p. 22.

5
the intellectual history of the middle ages

mishtooni bose

According to Alain de Libera, 'the thirteenth-century appearance of the intellectual' ('*l'apparition de l'intellectuel*') was a decisive moment in the history of the West.[1] Tracing the evolution of medieval intellectual life has led to similarly decisive developments in modern historiography. The study of medieval intellectual history is a particularly engaging and seductive task for modern scholars because the objects of our study can seem rather gratifyingly like ourselves: *clercs*, mediators between different social worlds, leading lives fraught with paradox; at once closely bound up with, and distant from, the worlds of commerce and politics, simultaneously admired and distrusted, consulted and controlled. The parallels could be taken further, casting modern intellectual historians as hopeful interpreters of culture, and thereby playing roles analogous to those of the medieval theologians described by Jean Dunbabin as 'go-betweens' in their own 'two-culture society'.[2]

It is the purpose of this chapter to map some of the interpretations of medieval intellectual culture that have been produced by such scholars over the last 50 years, to identify new directions for this field and to reappraise the usefulness of the term 'the history of ideas' in relation to it. For this period, as for others, such interpretations have many possible forms, and a multiplicity of scholarly narratives clamours for inclusion under the rubric 'The intellectual history of the Middle Ages', notwithstanding the fact that many scholars are reluctant to define themselves primarily as intellectual historians. Thus, our task often becomes one of diagnosing cases of intellectual history retrospectively, seeing what the dividends of particular scholarly narratives might be for our understanding of ideas in context. I will outline some of the major narratives here, firstly in the interests of the reader finding his or her way in this field for the first time, and, secondly, as a means of addressing the principal methodological question facing us – what might 'intellectual history' mean in relation to the study of the Middle Ages? Is it the 'history of intellectuals' or the 'history of ideas', or both?

The intellectual history of the Middle Ages is a field of study whose centre is everywhere and whose circumference is, apparently, nowhere: as for other periods, it is difficult, and quite possibly undesirable, absolutely to distinguish between 'intellectual' and other genres of medieval history, since, like the phenomena they analyse, social, political, institutional and intellectual histories are naturally part of one another, and many subsidiary areas not given substantial consideration here have important implications for our understanding of the intellectual cultures of the Middle Ages.[3] It might be legitimate for a chapter of this kind to focus entirely on the relationship between political and intellectual history, for example, since the close relationships between 'intellectual' and 'political' cultures, and their reciprocal influences during this period, are everywhere evident.[4] Later in this chapter, I will discuss in broader terms the way in which medieval intellectual history can trace the progress of ideas through different textual formations and under various social and political conditions. The historical period in question offered a variety of arenas for the metabolising, transformation and, above all, the vernacularisation of ideas: as Jacques Verger has pointed out, the 'global tendency' in the medieval centuries, is towards 'laicisation', a development analogous to the 'deprofessionalisation' of intellectual life described and analysed by de Libera.[5] In the first instance, however, and particularly if one is approaching the field for the first time, it is convenient to see the intellectual history of this period as suspended between two poles: the history of education, and of intellectual freedom. With the caveat above regarding the scope of the present study in mind, therefore, I will first give an overview of some of the areas of research that have contributed to our understanding and analysis of medieval intellectual life: the history of universities, of scholasticism, intellectual freedom and the contestation of religious orthodoxy.

i

In eavesdropping on snatches of medieval conversations from a privileged but ineluctably marginal position, modern intellectual historians have a vivid precursor in the form of an adopted Parisian, Christine de Pizan, in her many literary self-representations. In the second book of *Ladvision Christine*, she experiences Paris as *'la cité d'Athenes'*, where the advancement of learning is achieved through academic disputation.[6] Imagination allows her to penetrate *intra muros* and thus vicariously to experience this inherently disputatious culture: *'m'arrestoie entre les escoliers de diverses facultez de sciences disputans ensemble, de maintes questions fourmans plusieurs argumens'*.[7] The vantage-point that she adopts here is that of the perennially extramural writer who is nevertheless privileged to observe life in the thick of academic work (*'entre les estudes'*). Christine's perspective, the implications of which will be discussed in more detail later in this chapter, provides a singularly ironic point of entry into our survey of medieval intellectual history. For now, the focus on Paris brings us at once to the seminal influence of Jacques Le Goff's 1957 study, *Les intellectuels au Moyen Age*, which could plausibly be credited with having invented 'medieval intellectual history' as a distinct and self-conscious field of study.[8] Nevertheless, Le Goff restricted his attention to the *'maîtres d'écoles'* (schoolmen), whose work was thought and the teaching of thought (*'ceux qui font métier de penser et d'enseigner leur pensée'*).[9] In the introduction to the book's first edition, he was confidently exclusive, warning his readers that they would not find there any sustained consideration of a number of arguably important groups of *savants*: cloistered mystics, poets or chroniclers from the extramural world (*'éloignées du monde des écoles'*). Most significantly, his was a definition of medieval intellectual life that explicitly excluded any recognition of the contribution of humanism to the development of traditions of thought: for Le Goff in 1957, the *savant* of the Renaissance was quite opposed to his preferred object of study (*'[il] s'oppose précisément à l'intellectual médiéval'*).[10]

As will become clear later in this chapter, it is a measure of how far modern historiography has since travelled in pursuit of a more nuanced appreciation of medieval intellectual life that such a narrow characterisation of the medieval intellectual himself (the gendered term is deliberate, and raises further questions to be dealt with below) is now barely serviceable. Later, we will consider Le Goff's own revision of his initial position, together with other methodological considerations. But for now it suffices to return to the self-contained subject of his original study, since histories of medieval universities have formed much of the internal framework for medieval intellectual history in recent decades. This development is exemplified in the work of scholars such as Gordon Leff, Jacques Verger, Jeremy Catto and William Courtenay, all of whom have studied the evolution of ideas as affected by the organisation and evolution of medieval universities.[11] As a tributary of this large category of studies, others take the setting of particular medieval universities as their point of departure, but focus on particular episodes in the evolving relationship between institutional and intellectual life. Such studies are numerous, but could be exemplified by the work of Zénon Kaluza on the nominalist-realist controversies, Leon Baudry's collection and study of texts from the early years of the University of Louvain when that institution was grappling with the legacy of the Wycliffite controversies, and Hester Goodenough Gelber's study of Dominican philosophy in early fourteenth-century Oxford, which is a good recent example of a particularly situated intellectual history exploring the 'gaseous exchange' between ideas and context.[12] Such histories must also be supplemented by the many studies of other educational institutions beyond the universities which have, at the least, important implications for our understanding of the variety of settings in which ideas were formed, transformed and transmitted.[13]

In a second category, beyond these confluences of intellectual and institutional history, there are studies that are confined to analysing the characteristics and the internal evolution of 'scholasticism' as a distinct literary and intellectual culture (although 'confined' is scarcely an appropriate word given the scope and ambition of this rich body of work). These range from the early twentieth-century revalorisations of scholasticism by Martin Grabmann, Maurice de Wulf and Marie-Dominique Chenu, to the penetrating readings of scholastic psychology and exegesis by Odon Lottin and Henri de Lubac, and on to the later work of a spectrum of scholars, such as Norman Kretzmann, Mark D. Jordan, Allan Wolter and Katharine Tachau, who have traced the evolution of logic, theology and philosophy in the setting of the medieval universities, and the contribution of particular individuals to the evolution of theological doctrines.[14] Riccardo Quinto's examination of the medieval and post-medieval development and implications of the term *'scholastica'* likewise puts a body of suggestive material regarding scholastic self-consciousness at the disposal of the intellectual historian.[15]

Thirdly, moving beyond the schools, the relations between oral and literate cultures are clearly of paramount importance in affecting the development and dissemination of ideas. Brian Stock has furthered our understanding of the impact of literacy on the intellectual formation of early medieval scholars and the contestation of religious doctrines, and Paul Saenger has illuminated the development of silent reading in this period and its implications for intellectual life.[16] Stock's classic exploration of the way in which the growth of literacy in the early Middle Ages contributed to the development of 'heretical and reformist movements' has its counterpart in a collection of essays that collectively subject the relationship between heresy and literacy to a wide-ranging interrogation.[17] Studies such as these provide particularly clear examples of the way in which what we now call 'intellectual freedom' keeps surfacing as a natural and often central concern of intellectual historians of this period. This theme unites the last areas of study to be mentioned here.

In the fourth category are studies that address tensions within the intellectual world of Christendom and its creative or problematic relationships with other cultures. Gilbert Dahan, George Makdisi and Anna Sapir Abulafia have contributed to a rich seam of work exploring the relationships between Christian, Jewish and Arab intellectual worlds during this period.[18] Abulafia's studies of Christian-Jewish polemical disputations are but one example of the way in which modern scholarship has explored the potential of the dialogue, a fundamental genre of medieval discourse, linking oral and literate cultures, intra- and extramural worlds, and in particular its ability to bring into focus the consequences for the development of ideas of confrontation, whether real or imagined, between different cultures.[19] The work of R. I. Moore on persecution in medieval societies has its *antistrophe* in the work of Cary

Nederman on ideas of toleration in the same period.[20] William Courtenay, Heinrich Fichtenau, Mary M. McLaughlin and J. M. M. H. Thijssen have asked searching questions about the limits and nature of medieval intellectual freedom, and have answered these with subtlety, substance and finesse.[21] This in turn leads us to confrontations with heresy as the cultural matrices for the development and transformation of ideas, and in this area studies of the Cathars and Lollards are paradigmatic.[22]

It is also necessary to acknowledge here not only the work of individual scholars, but of the virtual communities constituted by the contributors to, and readers of, journals such as *Viator, Traditio, Medievalia et Humanistica, Vivarium* and the *Revue des sciences philosophiques et théologiques* in fostering the development of intellectual history. The grand narratives of individual scholars have not merely been complemented but enabled, and even driven, by the work of others such as Olga Weijers and Jacqueline Hamesse, who have presided over what amount to modern *ateliers* of intellectual artisans (as in the studies published by Brepols in the *Studia Artistarum* series) reconstructing the physical and mental apparatus, the *instruments de travail*, of the medieval scholar.[23] This essential collaborative work, no less than the critical syntheses of individual scholars, has substantially enriched our understanding of the quotidian life of the medieval universities and the material conditions, the codicological organisation and the literary genres that served the development, contestation and dissemination of ideas. We have already briefly noted the fundamental importance of medieval dialogues and disputations.[24] Another paradigmatic literary genre is the *Speculum Principis*, the 'Mirror for Princes', the medieval history of which provides a substantial example of the relationship between, on the one hand, the evolution of ideas and texts and, on the other, the evolution of the political communities that they variously interpellate.[25]

ii

As the discussion of Le Goff above indicates, the non-Anglophone world has been, perhaps predictably, rather more concerned than has the Anglophone with the apparent difficulty of attaining a satisfactory application of the self-consciously anachronistic terms 'intellectual' and 'intellectual history' to the description and analysis of medieval cultural, social and institutional phenomena. Anglophone scholarship is comparatively insouciant about this matter. This attitude is exhibited in Marcia Colish's recent *grands récits*, which appraise the broader trajectories of twentieth-century contributions to medieval intellectual history. Colish, the only scholar defined as an 'intellectual historian' in a recent collection of biographical essays honouring the contributions of women to medieval studies, has not only contributed substantial and seminal studies that have developed the field further, but has also provided the means with which to undertake further intellectual cartography of the period.[26] If Le Goff is, in Dunbabin's words, 'the most Platonic of historians', Colish is an Aristotelian, her attitude to the subject embracing and emphasising 'conjunctures, changes, turning-points, ruptures and differences'. [27] For her, 'intellectuals' and 'intellectual history' are flexible but essentially self-explanatory terms. While emphasising the importance of patronage networks in this period, she is convinced that these did not compromise the intellectual freedom of medieval academics, whose 'education taught them that there was more than one way to think about the same subject, and that subjects could be taught in diverse ways within the orthodox consensus, even in the high-risk field of theology'. [28] In her 2000 Etienne Gilson Lecture, *Remapping Scholasticism*, she confronts and marginalises what she represents as 'two competing story lines for medieval intellectual history', both of which evolved in the early twentieth century: the first an 'anti-Burckhardtian effort ... to push Renaissance humanism ... into the Middle Ages' (as exemplified in Charles Homer Haskins's 1927 study *The Renaissance of the Twelfth Century*), and the second a 'neo-Thomist revival' launched by Roman Catholic scholars (such as De Wulf, Grabmann, Lottin and Jacques Maritain) concerned to prove the 'intellectual respectability of Catholic philosophy'.[29] Colish shows how more recent scholarship has complicated the trajectories of both narratives, involving a 'decentring of Aquinas' in favour of attention to Scotus and Ockham, and the uncoupling of confessionalism from scholarship, as exemplified in the anti-scholastic emphases of Jean Leclerq and the historicism of Chenu, R. W. Southern and Giles Constable.[30] She describes how modern narratives have become preoccupied with delineating the developments in scholastic thought regarding God's omnipotence and issues of divine foreknowledge, free will and ethics, and she summarises the advancements of modern learning concerning medieval logic and philosophy. She ends by enjoining on current and future scholars the study of philosophy and theology together and ends by noting the salutary fact that 'current scholarship has committed itself to recovering the multiple itineraries through the scholastic landscape that actually existed in the medieval period itself'.[31]

The promise of multiplicity is pursued elsewhere in Colish's work. In the synthesis which could be regarded as the culmination of her contribution to the intellectual history of the Middle Ages, she brings together her diverse subjects under the umbrella term 'Christian intellectuals'.[32] Nevertheless, in the other methodological essays that have overlapped with the writing of this work, she has identified 'Roman Catholic triumphalism' as a predominant strain in earlier twentieth-century contributions to medieval intellectual history, and points out that '"Christianization" meant different things to different people in the Middle Ages'.[33] When pondering the future of the sub-discipline, and the strategies that it might adopt in order to ensure its survival, she recommends that modern historians should be alert to the possibilities of 'a new story line for medieval intellectual history ... the story of how a European tradition was formed in the first place, out of the interactions among indigenes and the bearers of cultural traditions that came, originally, from outside of the European landmass'. A way is thereby opened for medieval intellectual history of the future to be the analysis of the series of dynamic processes whereby 'this traditional society made of its past ... not a sterile or static cultural icon but a catalyst, promoting change and modernization from within'.[34] Most importantly, Colish's reason for urging this course of action on her colleagues is not cynical (a weary capitulation to political correctness), or even merely pragmatic (a means of ensuring the institutional survival of the sub-discipline): rather, they are enjoined upon us all because in her view such a perspective provides fresh access routes to new truths about this period.

iii

Although Colish's own work concentrates on scholasticism as traditionally conceived – that is, the activity of schoolmen – one of the implications of her remarks concerning the diversity and multiplicity of routes through medieval thought is that we cannot view 'intellectuals' as being the exclusively intramural species analysed so influentially in Les intellectuels au Moyen Age. Le Goff's own modifications, in the introduction to the 1985 edition of Les intellectuels, to the terms of reference operative in the original study exemplify the way in which modern historiography has sought to attune itself to the ever-widening spectrum of appropriate and analysable phenomena in this field of study.[35] The 1957 text is unaltered and he retains some of its emphases, notably on men rather than institutions and social structures rather than ideas per se, but the general trend of his thought in this introduction is towards a broadening of the original edition's horizons. He emphasises the critical dimension of scholasticism, seen in cases of heresy and condemnation (Abelard, Aquinas, Siger de Brabant and Wyclif are acknowledged). He acknowledges the fundamental work of Peter Lombard and Peter Comestor. More significantly still, he repents of not having included some discussion of Dante and Chaucer in the earlier edition, finally acknowledges (it is tempting to say 'admits') that there are traces of both humanism and scholasticism at work in the thought of later schoolmen, and names Jean Gerson and Nicholas of Cusa as exemplary in this respect. This recognition is made en route to the broader realization that the assumption of a fundamental opposition between humanism and scholasticism, one that severely limits the scope and impact of his thesis, requires revision.[36]

Le Goff's self-criticism anticipates the conclusion of the present essay, which seeks to broaden the terms of reference under which we study 'the medieval intellectual' still further. Other studies point in the same direction. It is inconceivable, for instance, that we should exclude the various phases and manifestations of medieval humanism (however we define that phenomenon) from intellectual histories of this period.[37] The abiding influence of Les intellectuels, together with the clear limitations of that study, are both acknowledged in de Libera's Penser au Moyen Age (1991), which takes the definition of 'the intellectual' in a different direction. De Libera begins with an acknowledgement of Le Goff's influence, but moves quickly on, taking it as having been demonstrated that a group of 'intellectuals' existed in the Middle Ages. Unlike Le Goff, who concentrated principally on theologians, de Libera is concerned with philosophy and its problematic place in medieval intellectual life, particularly its fraught relationship with theology. In particular, he seeks to delineate the 'deprofessionalisation of philosophy', for him a crucial moment in medieval history ('*la déprofessionnalisation de la philosophie est donc ... ce qui signe le veritable moment de la naissance des intellectuels*').[38] This leads naturally to the consideration of groups and individuals beyond the self-limiting purview of Le Goff's study: those who incarnate a general tendency fundamental to the evolution of medieval thought: the laicisation of thought and the translation of important texts into the vernacular ('*la laicisation de la pensée et le passage à la langue vernaculaire*').[39] The mediators of the philosophical ideal, he argues, include figures such as Dante and Meister Eckhart, figures marginal to Le Goff's original terms of reference. De Libera's chapter on 'Philosophes et intellectuels' insists on the importance of what medieval intellectual life in the West inherited from Arabic thought: not only texts, but tensions, which were replicated in the oppositions between philosophy and Christian theology. De Libera's narrative, furthermore, includes both conventional schoolmen and the 'philosopher-kings', Emperor Frederick II and his son Manfred, king of Sicily. His is, therefore, not merely an account of the vernacularisation of thought, but a provocative enactment of it, moving away from the resolutely intramural focus of Le Goff and entailing a redefinition and broadening of what the term 'philosopher' might be made to mean as a category for analysis.

In questioning the latitude of the term 'philosopher', de Libera broaches an important issue, namely the extent to which 'intellectual history' was made as decisively in the world beyond the medieval schools as it was within them. This is an approach adumbrated by Giovanni Tabacco, who has examined the relationships between intellectuals and power structures in the Middle Ages and defends the applicability of the term 'intellectuals' to figures such as priests, ascetics and jurists whose roles were not primarily pedagogic, but who assumed responsibility for 'rationalising' systems of thought in relation to fundamental issues such as the limits of papal power and relations between secular *imperium* and ecclesiastical hierarchy.[40] Tabacco traces these processes back to late antiquity, drawing together events such as the Council of Nicaea, individuals such as Cassiodorus and Gregory the Great, and texts such as the False Decretals of Pseudo-Isidore in a rich analytical narrative that demonstrates the profound interdependence of political and intellectual history. There is a degree of continuity between Tabacco's identification of Nicaea as an event in intellectual history and the approach of Michael Shank, as seen in a rightly influential study that shows how arguments at the Council of Constance concerning the use of logic in theology had important consequences for the shaping of intellectual life both within and beyond the late-medieval schools.[41] Such concern for the increasingly public discourse in which late-medieval theologians were engaged is also reflected in the recent work of Maarten Hoenen and Ian Levy on the terms on which one of John Wyclif's early opponents confronted the intimations of heterodoxy in his arguments.[42]

Like that of de Libera, such intellectual histories situate themselves at the boundaries between intra- and extramural worlds. This is a significant characteristic in recent studies of late-medieval intellectual life, and thus such developments, together with Le Goff's belated revaluation of Gerson and Cusa as intellectuals, bring us to the consideration of new narratives about intellectual life in the fifteenth century. A recent collection of essays addressing Johan Huizinga's influential portrait of France and the Netherlands in this period grapples with the distorted perspective on this period unwittingly sponsored by the English and French titles of Huizinga's most famous work and returns us to consideration of the fifteenth century as a 'harvest', a gathering-in and independently fruitful sifting of traditions of thought.[43] These essays show how a range of analyses can take as their point of departure traditional categories, such as the history of universities, of academic disciplines and particular individuals, and use them to effect a collectively strong rereading of a notoriously complex period in intellectual history.[44] The structure of the volume singles out Gerson and Cusa for particular attention, and Gerson's influence on the intellectual character of this period is acknowledged in essays beyond those in the section explicitly allotted to him.[45]

Daniel Hobbins's recent work on Jean Gerson focuses many of these issues in his conception of the late-medieval theologian as a 'public intellectual'.[46] This is, in part, a reformulation of the established lines of enquiry that we have been tracing, but it refreshes them nonetheless, enabling us to think further about the transformation of ideas, the literary genres in which they were disseminated in the worlds beyond the schools, and the careers of the disseminators. Hobbins's essay and dissertation partly consolidate and partly originate new ways of thinking about the intellectual. Particularly notable for our concerns here is the way in which the place he negotiates for his subject, Gerson, is dependent on sidestepping what he calls 'the history of ideas, in which Gerson will nearly always appear at best as a popularizer, at worst as a derivative and unoriginal theologian working in the twilight of a better age'. Instead, Hobbins announces his desire 'to reorient the discussion by turning the focus toward what may be called the "cultural positioning" of the late medieval schoolman, by which I mean his public status, his literary connection to a wider public, and hence his cultural relevance'.[47] The term 'public' is crucial here, connecting Hobbins's paradigm-shifting approach to Gerson with other studies of late-medieval controversies and controversialists in which intellectuals were engaged, with increasing specificity, in interpellating a public, or variety of publics, real and imagined, learned and unlearned, in literary genres derived from, but not circumscribed by, the demands of conventional academic exercises.[48] This crystallises in the centrality to Hobbins's thesis of the late-medieval tract, a genre that made ideas more portable and thus facilitated their transfusion.

iv

Hobbins's essay provoked a response complaining that, in keeping with American scholars' general lack of interest in the histories of Central and Eastern Europe, he had underestimated the contribution of Jan Hus to the evolution of the late-medieval intellectual life.[49] His good-tempered reply insisted that Gerson should be viewed as a paradigmatic rather than unique figure in his mapping of this territory. Auto-critique, however, remains necessary, since one potentially debilitating fear attendant on the enterprise of writing medieval intellectual history is that of unwitting parochialism. Nevertheless, as Marcia Colish has shown, there are many routes towards the broadening of horizons, and thus we return to one of the 'salient innovations' that she has noticed taking place in histories of this period: recognition of the influence of women on the character of its intellectual life.[50] Such recognition is, of course, hardly unproblematic, and thus it is hardly universal. Hobbins's necessarily selective appendix of medieval 'public intellectuals', for example, makes no mention of female authors (however liberally one defines 'authors' in this context) and one might deduce from it that women exercised no significant influence on the development of thought in this period. Nevertheless, a robust case can be made for taking into account the contribution of women to the formation of its public spheres of intellectual discourse, for reasons that effortlessly transcend the banalities and pseudo-exigencies of political correctness. Writing women into intellectual as well as social and cultural history simply reflects their contribution to the shared enterprises of interpellating and shaping readerships, and to the transformation ('vernacularisation' in both literal and metaphorical senses) of ideas. This can be seen in many places in late-medieval writing, and does not require that we recruit medieval writers to the ranks of modern feminisms. Chaucer, for example, repeatedly uses female characters as a means of exposing the contingencies of textual and intellectual history that lie behind the construction of exegetical and cultural hegemonies, but the characters through which he articulates such exposés are granted varying levels of awareness that there is a critical agenda at stake.[51]

Consideration of the role of women in medieval intellectual history returns us to Christine de Pizan. In Jacques Verger's *Les gens de savoir dans l'Europe à la fin du Moyen Age* (tellingly translated into English as *Men of Learning in Europe at the End of the Middle Ages*), Christine appears fleetingly as a 'semi-professional author', one of the 'rare women' whose familiarity with learned disciplines was such that one could number her among the learned.[52] Overtly feminist scholarship has converged with other genres of history, however, to render her increasingly intelligible to us.[53] Moreover, Christine's writings and intellectual engagements potently focus a number of other issues, notably the relationship between gender, heresy and intellectual emancipation. Christine's writing obliges us to think about the ways in which gender wrote its way into intellectual questions important to this period and, in certain cases, substantially affected the shape and transmission of ideas. Her intellectual relationship with Jean Gerson, exemplified as much in the congruence between their instinctively allegorising, Boethian imaginations as in her reactions to Gerson's ideas, is a provocative and focused example of the process of exchange between intra- and extramural worlds.[54] In what is arguably Jean Gerson's most well-known excursion into the vernacular, the treatise 'Against the Roman de la Rose' (18 May 1402), written in support of Christine's attack on the *Roman de la Rose*, the terms in which he presents the defence of Chastity are specifically those in which one might also undertake a defence of the true faith. There are parallels here with Christine's own *Livre de la Cité des Dames*, written around three years later, in which the literary traditions of misogyny, which the book seeks to displace, are associated with not merely degenerate but, more precisely, heretical thinking. If, therefore, medieval intellectual life may fruitfully be described in terms of progressive 'laicisation' or 'deprofessionalisation', it is not merely permissible but necessary to see such a process culminating in a further phase, one in which questions relating to the gendering of intellectual arenas determined the transformation, and in some cases the revision, of ideas. It could, of course, be legitimately argued, in keeping with Verger's comment, that Christine is an exceptional case. Against such an argument, however, one could posit the many late-medieval examples of women's religious experience (and often the written transmission of that experience) challenging the latitudes of orthodox theology and exerting a direct influence on its construction.[55]

v

Alexander Murray's study, *Reason and Society in the Middle Ages*, was an attempt to challenge a view of the medieval period as an uncontested 'age of faith', to expose the 'variety of traditions, both co-operating and competing' behind 'the monolithic appearance of the medieval church', and to do so by exploring the dialectic between tendencies of thought, both 'miraculous and rationalistic'.[56] Although Murray's study could be classified in various ways – it has implications for social, ecclesiastical and religious histories of the period – his title is instructive, recalling us to the fact that intellectual history need not be obliged to take 'intellectuals', however tightly or loosely defined, as its starting point. It can instead trace the life of ideas, concepts and habits of thought in particular social milieux. I thus conclude on what might seem to be a wilfully old-fashioned note, since phrases such as 'the life of ideas' may inadvertently recall older concepts and methods, in an apparent elision of 'intellectual history' with the 'history of ideas'. The return to ideas may also seem wilfully to reverse Le Goff's emphasis on individuals. But I contend that 'the history of ideas' remains not merely a potent phrase but a potent practice, as Alain Boureau implicitly recognises in his observation that 'the question of the social commitment of intellectuals and their strategies of engagement in the world ... has opened a path to a history of ideas'. [57] What is thereby enjoined on us is not some selective history of deracinated, disembodied, dematerialised or even idealised concepts, but an intellectual history conducted in the spirit of Gabrielle Spiegel's observation that 'texts, as material embodiments of situated language use, reflect in their very materiality the inseparability of material and discursive practices'. [58] A materialist intellectual history endeavours to trace the formation and transmission of ideas within and across different milieux, always with particular, situated human beings as the agents of such processes but also with other cultural continuums (rather than oppositions) between clerisy and laity, orality and literacy, Latinity and vernacularity, female and male gendered roles, borne constantly in mind. It is thus that we can trace the changing currency, implication and resonance of ideas as they pass from the schools to the world beyond, from biblical exegesis to vernacular sermons or political treatises. Extending the territory of intellectual history in the ways I have described above, taking it beyond the medieval schools, and observing it being made wherever 'ideas' are in question, only furthers this process, since it obliges us to consider the transformation of ideas that occurs whenever and wherever cultural and institutional fault lines are crossed, and *translationes* of various kinds are set in motion.

notes

1. Alain de Libera, *Penser au Moyen Age* (Paris, 1991), p. 143.
2. Jean Dunabin, 'Jacques Le Goff and the Intellectuals' in Miri Rubin, ed., *The Work of Jacques Le Goff and the Challenges of Medieval History* (Woodbridge, 1997), pp. 157–67 (p. 165).
3. Those studying this field for the first time are directed in the first instance to three further sources: Alain Boureau, 'Intellectuals in the Middle Ages, 1957–1995' in Rubin, *The Work of Jacques Le Goff*, pp. 145–55; the work of Marcia Colish, and in particular the bibliography that can be conveniently reconstructed from her essay *Remapping Scholasticism*, The Etiènne Gilson Series (Toronto, 2000) (discussed below); and Daniel Hobbins, 'The Schoolman as Public Intellectual: Jean Gerson and the late medieval tract', *American Historical Review* 108 (2003), 1308–37 (also discussed below). In such a rich field, and within a comparatively short space, it seems more than usually invidious to single out the work of certain individuals at the expense of that of others. Accordingly, in what follows, the bibliographical references given here should be seen as indicative rather than exhaustive.
4. One example of a study that demonstrates the seamless relationship between the intellectual and political history of this period is Philippe Buc, *L'ambiguité du livre: prince, pouvoir et peuple dans les commentaires de la Bible au Moyen Age* (Paris, 1994). Another example might be Ernst Kantorowicz, *The King's Two Bodies: a study in mediaeval political theology* (Princeton, NJ, 1957). Brian Tierney's studies of medieval

political thought have obvious dividends for intellectual history: see, for example, *Foundations of the Conciliar Theory: the contribution of the medieval canonists from Gratian to the Great Schism* (Cambridge, 1955). The relationship between purveyors of ideas and political power in this period is analysed by Giovanni Tabacco, 'Gli intellettuali nel medioevo nel givoco delle instituzione edelle preponderanze sociali' in Corrado Vivanti, ed., *Storia d'Italia. Annali 4: Intellettuali e Potere* (Turin, 1981), pp. 5–46 (discussed further below).

5. Jacques Verger, *Les gens de savoir dans Europe à la fin du Moyen Age* (Paris, 1997), p. 207. An English version of this work has appeared as *Men of Learning in Europe at the End of the Middle Ages*, trans. Lisa Neal and Steven Randall (Notre Dame, Ind., 2000). For de Libera on 'deprofessionalisation' (*Penser au Moyen Age*, pp. 12–13), see further below.

6. Christine de Pizan, *Le livre de l'advision Christine*, eds, Christine Reno and Liliane Dulac (Paris, 2001), p. 51. An English translation is available as *The Vision of Christine de Pizan*, trans. Glenda McLeod and Charity Cannon Willard (Cambridge, 2005).

7. Ibid.

8. Jacques Le Goff, *Les intellectuels au Moyen Age* (Paris, 1957, 1960, 1985).

9. *Les intellectuels au Moyen Age* (Paris, 1957), p. 4.

10. Ibid.

11. Gordon Leff, *Paris and Oxford Universities in the Thirteenth and Fourteenth Centuries: an institutional and intellectual history* (New York and London, 1968); Jacques Verger, ed., *Histoire des universités en France* (Toulouse, 1986), and 'Histoire intellectuel' in Michel Balard, ed., *L'Histoire médiévale en France: bilan et perspectives* (Paris, 1991), pp. 177–97; Jeremy Catto, 'Wyclif and Wycliffism at Oxford 1356–1430' and 'Theology after Wycliffism' in J. I. Catto and R. Evans, eds, *The History of the University of Oxford*, Vol. 2: *Late medieval Oxford* (Oxford, 1992), pp. 175–261, 263–80; William J. Courtenay, *Schools and Scholars in Fourteenth-century England* (Princeton , NJ, 1987). Verger and Courtenay are prolific in this area and their many articles should be consulted. See further H. De Ridder-Symoens, ed., *A History of the University in Europe*, Vol. 1: *Universities in the Middle Ages* (Cambridge, 1992), in particular chapters 5 and 13, and the epilogue, in which Walter Rüegg considers the impact of early humanism on and within the medieval university. Other studies that take the history of universities as their point of departure, but which focus on areas important to the development and transmission of ideas, include John Van Engen, ed., *Learning Institutionalized: teaching in the medieval university* (Notre Dame, Ind., 2000); Alexander Patschovsky and Horst Rabe, eds, *Die Universität in Alteuropa* (Constance, 1994); James M. Kittelson and Pamela J. Transue, eds, *Rebirth, Reform and Resilience. Universities in transition 1300–1700* (Columbus, Ohio, 1984); Maarten J. F. M. Hoenen, J. H. Josef Schneider and Georg Wieland, eds, *Philosophy and Learning: universities in the Middle Ages* (Leiden, 1995); and Nancy Van Deusen, ed., *The Intellectual Climate of the Early University. Essays in honor of Otto Grundler* (Kalamazoo, Mich., 1997). In a particularly nuanced study, Ian P. Wei attempts to understand 'the nature of intellectual activity' and its relationship to 'the place or function of universities in medieval society as a whole' in 'The Masters of Theology at the University of Paris in the Late Thirteenth and Early Fourteenth Centuries: an authority beyond the schools', *Bulletin of the John Rylands Library* 75 (1993), 37–63.

12. Zénon Kaluza, *Les querelles doctrinales à Paris: nominalistes et réalistes aux confins du XIVe et du XVe siècles* (Bergamo, 1988); Zénon Kaluza and Paul Vignaux, eds, *Preuve*

et raisons à l'Université de Paris: logique, ontologie et théologie au XIVe siècle (Paris, 1984); Léon Baudry, ed., *La Querelle des Futurs Contingents (Louvain 1465–1475). Texts inédits* (Paris, 1950), translated by Rita Guerlac as *The Quarrel Over Future Contingents* (Dordrecht, Boston, Mass., and London, 1989); Hester Goodenough Gelber, *It Could Have Been Otherwise: contingency and necessity in Dominican theology at Oxford, 1300–1350* (Leiden, 2004).

13. See, for example, C. Stephen Jaeger, *The Envy of Angels: cathedral schools and social ideals in medieval Europe, 950–1200* (Philadelphia, Penn., 1994); Bert Roest, *A History of Franciscan Education (c. 1210–1517)* (Leiden, 2000).

14. Martin Grabmann, *Die Geschichte der scholastichen Methode* (Freiburg, 1909–11); Maurice de Wulf, *Histoire de la Philosophie Médiévale* (6th edn, Louvain, 1937–47); M.-D. Chenu, *La théologie comme science au XIIIe siècle* (3rd edn, Paris, 1957): D. Odon Lottin, *Psychologie et Morale au XII et XIII siècles* (Louvain, 1942–60); Henri de Lubac, *Exegese médiévale: Les quatre sens de l'Ecriture* (Paris, 1959–64); Norman Kretzmann, Anthony Kenny and Jan Pinborg, eds, *The Cambridge History of Later Medieval Philosophy. From the rediscovery of Aristotle to the disintegration of scholasticism, 1100–1600* (Cambridge, 1982); Mark D. Jordan, *Ordering Wisdom: the hierarchy of philosophical discourses in Aquinas* (Notre Dame, Ind., 1986); Allan B. Wolter, *The Philosophical Theology of John Duns Scotus*, ed. Marilyn McCord Adams (Ithaca, NY, and London, 1990); Katherine H. Tachau, *Vision and Certitude in the Age of Ockham: optics, epistemology and the foundations of semantics, 1250–1345* (Leiden, 1988). For further references and discussion, see Colish, *Remapping Scholasticism*.

15. Riccardo Quinto, '"Scholastica". Contributo alla storia di un concetto', *Medioevo* 17 (1991), 1–82; 'II: Secoli XIII–XVI', 19 (1993), 67–166; 'III: Dal XVI al XVIII secolo', 22 (1996), 335–450.

16. Brian Stock, *The Implications of Literacy: written language and models of interpretation in the eleventh and twelfth centuries* (Princeton, NJ, 1983); Paul Saenger, *Space Between Words: the origins of silent reading* (Stanford, Calif., 1997). A classic early study is Herbert Grundmann, 'Literatus - Illiteratus. Der Wandel einer Bildungsnorm vom Altertum zum Mittelalter', *Archiv fur Kulturgeschichte* 40 (1958), 1–65. A more recent collection of essays analysing the 'implications of literacy' in a different setting at the close of the Middle Ages is Monique Ornato and Nicole Pons, eds, *Pratiques de la culture ecrite en France au XV* siècle* (Louvain-la-Neuve, 1995).

17. Stock, *Implications of Literacy*, pp. 145–51, 241–325; Peter Biller and Anne Hudson, eds, *Heresy and Literacy 1100–1530* (Cambridge, 1994).

18. Gilbert Dahan, *Les intellectuels chrétiens et les juifs au Moyen Age* (Paris, 1990); George Makdisi, *The Rise of Humanism in Classical Islam and the Christian West: with special reference to scholasticism* (Edinburgh, 1990); Anna Sapir Abulafia, *Christians and Jews in the Twelfth-Century Renaissance* (London and New York, 1995).

19. Anna Sapir Abulafia, *Christians and Jews in Dispute: Disputational Literature and the Rise of Anti-Judaism in the West (c. 1000–1150)* (Aldershot, 1998).

20. R. I. Moore, *The Formation of a Persecuting Society: power and deviance in Western Europe, 950–1250* (Oxford, 1987); John Christian Laursen and Cary Nederman, eds, *Beyond the Persecuting Society: religious toleration before the Enlightenment* (Philadelphia, Penn., 1998); Cary Nederman, *Worlds of Difference: European Discourses of Toleration, c. 1100–c. 1500* (Philadelphia, Penn., 2000).

21. William Courtenay, 'Inquiry and Inquisition: academic freedom in medieval universities', *Church History* 58 (1989), 168–181, and 'Dominicans and Suspect Opinion in the Thirteenth Century: the cases of Stephen of Venizy, Peter of Tarentaise, and the articles of 1270 and 1271', *Vivarium* 32 (1994), 186–95; Heinrich Fichtenau,

Ketzer und Professoren: Häresie und Vernunftglaube im Hochmittelalter (Munich, 1992), translated as *Heretics and Scholars in the High Middle Ages, 1000–1200*, trans. Denise A. Kaiser (University Park, Penn., 1998); Mary M. McLaughlin, *Intellectual Freedom and its Limitations in the University of Paris in the Thirteenth and Fourteenth Centuries* (New York, 1977); J. M. M. H. Thijssen, *Censure and Heresy at the University of Paris, 1200–1400* (Philadelphia, Penn., 1998).

22. Malcolm Barber, *The Cathars: dualist heretics in Languedoc in the High Middle Ages* (Harlow, 2000); John Arnold, *Inquisition and Power: Catharism and the confessing subject in medieval Languedoc* (Philadelphia, Penn., 2001); Anne Hudson, *The Premature Reformation. Wycliffite texts and Lollard history* (Oxford, 1988), and *Lollards and Their Books* (London, 1985); Margaret Aston, *Lollards and Reformers: images and literacy in late medieval religion* (London, 1984).

23. Both Hamesse and Weijers are prolific as authors, editors and facilitators of research, and the following are only indicative: Jacqueline Hamesse, ed., *Manuels, programmes de cours et techniques d'enseignement dans les universités médiévales* (Louvain-la-Neuve, 1994), and Jacqueline Hamesse, ed., *L'élaboration du vocabulaire philosophique au Moyen âge: actes du colloque international de Louvain-la-Neuve et Leuven, 12–14 Septembre 1998* (Turnhout, 2000); Olga Weijers, ed., *Le maniement du savoir: pratiques intellectuelles à l'époque des premières universités (XIIIe–XIVe siècles)* (Turnhout, 1996); Olga Weijers and Louis Holtz, eds, *L'enseignement des disciplines à la faculté des arts (Paris et Oxford, XIIIe–XVe siècles): actes du Colloque international* (Turnhout, 1997), and Olga Weijers, ed., *Le travail intellectuel a la Faculté des arts de Paris: textes et maitres (ca. 1200–1500)* (Turnhout, 1994). The Brepols series *Typologie des sources du Moyen Age Occidental* could also be singled out as having contributed substantially to various aspects of medieval intellectual history by consolidating the current state of learning regarding a broad range of medieval modes and genres of discourse (including sermons, disputations and letter-writing).

24. On the genre in its university setting, see, for example, *Les genres littéraires dans les sources théologiques et philosophiques médiévales: définition, critique et exploitation. Actes du colloque international de Louvain-la-Neuve, 25–27 mai 1981* (Louvain-la-Neuve, 1982); Bernardo C. Bazan, ed., *Les Questions disputées et les questions quodlibétiques dans les facultés de théologie, de droit et de medicine* (Turnhout, 1985); Olga Weijers, ed., *La 'disputatio' à la Faculté des arts de Paris (1200–1350 environ): esquisse d'une typologie* (Turnhout, 1995).

25. The place of one text in medieval intellectual life is traced in Charles F. Briggs, *Giles of Rome's De regimine principum: reading and writing politics at court and university, c.1275–c.1525* (Cambridge, 1999). Briggs focuses in particular on what can be deduced from the various codicological histories of the text, but the implications for comparative intellectual history are self-evident.

26. E. Anne Matter, 'Marcia Colish' in Jane Chance, ed., *Women Medievalists and the Academy* (Madison, Wisc., 2005), pp. 980–94.

27. Dunbabin, 'Jacques Le Goff and the Intellectuals', p. 157.

28. Marcia Colish, 'Re-envisioning the Middle Ages: a view from intellectual history' in Roger Dahood, ed., *The Future of the Middle Ages and the Renaissance. Problems, trends and opportunities for research* (Turnhout, 1998), pp. 19–26 (p. 24).

29. Colish, *Remapping Scholasticism*, pp. 1, 5.

30. Ibid., 8, 11.

31. Ibid., 21.

32. Marcia Colish, *Medieval Foundations of the Western Intellectual Tradition 400–1400* (New Haven, Conn., and London, 1997), p. 7.

33. Marcia Colish, 'Intellectual History' in John Van Engen, ed., *The Past and Future of Medieval Studies* (Notre Dame, Ind., and London, 1994), pp. 190–203 (p. 199). See also Colish, 'Re-envisioning the Middle Ages'.

34. Colish, 'Intellectual History', p. 197.

35. Le Goff, *Les intellectuels au Moyen Age* (Paris, 1985). See further Le Goff, 'Définition de l'intellectuel et ses implications' in Jacques Le Goff and Béla Köpeczi, eds., *Intellectuels Français, Intellectuels Hongrois XIIIe-XXe Siècles* (Budapest and Paris, 1985), pp. 11–23.

36. Le Goff, *Les intellectuels au Moyen Age* (1985), pp. 1, 3, 4, 9.

37. R. W. Southern, for example, has fruitfully explored aspects of intellectual life in early and high medieval Europe, tracking the movement of ideas in and beyond centres of learning, in *Scholastic Humanism and the Unification of Europe*. Vol. 1: *Foundations* (Oxford, 1995); Vol. 2: *The Heroic Age*, with notes and additions by Lesley Smith and Benedicta Ward (Oxford, 2000). The range of studies of fifteenth-century humanism is considerable, but James Hankins, *Plato in the Italian Renaissance* (Leiden, 1990), is an excellent representative study.

38. De Libera, *Penser au Moyen Age*, pp. 12–13.

39. Ibid., p. 13. On the teaching of French in England and France during this period, and its implications for intellectual life, see Serge Lusignan, *Parler Vulgairement. Les intellectuals et la Langue Française aux XIIIe et XIVe Siècles* (Paris and Montreal, 1987). See further Rita Copeland, *Rhetoric, Hermeneutics and Translation in the Middle Ages: academic traditions and vernacular texts* (Cambridge, 1991).

40. Tabacco, 'Gli intellettuali nel medioevo nel giuoco delle istituzione e delle preponderanze sociali', p. 11.

41. Michael Shank, *'Unless You Believe, You Shall Not Understand'*: *logic, university and society in late-medieval Vienna* (Princeton, NJ, 1988).

42. Maarten J. F. M. Hoenen, 'Theology and Metaphysics. The debate between John Wyclif and John Kenningham on the principles of reading the Scriptures' in Mariateresa Fumagalli Beonio Brocchieri and Stefano Simonetta, eds, *John Wyclif. Logica, politica, teologia. Atti de Convegno Internazionale Milano, 12–13 Febbraio 1999* (Florence, 2003), pp. 23–55; Ian Levy, 'Defining the Responsibility of the Late Medieval Theologian: the debate between John Kynyngham and John Wyclif', *Carmelus* 49 (2002), 5–29.

43. Jan A. Aertsen and Martin Pickavé , eds, *'Herbst des Mittelalters?': Fragen zur Bewertung des 14. und 15. Jahrhunderts* (Berlin, 2004). Johan Huizinga's study, *Herfsttij der Middeleeuwen* (1919), was translated into English by F. Hopman as *The Waning of the Middle Ages* and into French as *Le déclin du Moyen Age*.

44. See, in particular, William Courtenay, 'Huizinga's Heirs: interpreting the Late Middle Ages' in Aertsen and Pickavé , *'Herbst des Mittelalters?'*, pp. 25–36. Courtenay is candid about the distance between Huizinga's confident but selective characterisation of late-medieval mentalities and the materials and methods at work in modern intellectual histories of the same period.

45. See, for example, Maarten J. F. M. Hoenen, 'Zurück zu Autorität und Tradition. Geistesgeschichtliche Hintergründe des Traditionalismus an den spätmittelalterlichen Universitäten' in Aertsen and Pickavé, *'Herbst des Mittelalters?'*, pp. 133–46, a vintage example of the confluence between institutional and intellectual history, covering a wide range of issues: the character of late-medieval Aristotelianism, the confrontation between 'nominalists' and 'realists' in the fifteenth century, the aftermath of the Wycliffite controversies and their influence on view of the relationship between philosophy and theology.

46. Daniel Hobbins, 'Beyond the Schools: new writings and the social imagination of Jean Gerson' (unpublished PhD dissertation, University of Notre Dame, 2002); 'The Schoolman as Public Intellectual, 1308–37.

47. Hobbins, 'The Schoolman as Public Intellectual', p. 1309.

48. Jeremy Catto couples Gerson with 'Wyclif and his intellectual contemporaries' as members of 'a new and more broad-minded elite, whose knowledge was readily accessible to the educated laity. They would have been more at home in the age of Erasmus than in that of Henry of Ghent' ('Wyclif and Wycliffism', pp. 218–19). Again, the trajectory described here is one of the gradual 'laicisation' of thought.

49. M. Mark Stolarik, 'Communication', *American Historical Review* 109 (2004), x.

50. Colish, 'Re-envisioning the Middle Ages', p. 20.

51. For one rich discussion of this tendency, see Ralph Hanna III, '*Compilatio* and the Wife of Bath: Latin backgrounds, Ricardian texts' in R. Hanna III, *Pursuing History. Middle English manuscripts and their texts* (Stanford, Calif., 1996), pp. 247–58.

52. Verger, *Les gens de savoir*, p. 107.

53. Kate Langdon Forhan, *The Political Theory of Christine de Pizan* (Aldershot, 2002); Eric Hicks, ed., with Diego Gonzalez and Philippe Simon *Au champ des escriptures: Actes du IIIᵉ colloque international sur de Christine de Pizan* (Paris, 2000); Angus J. Kennedy et al., eds, *Contexts and Continuities: proceedings of the IVth International Colloquium on Christine de Pizan, Glasgow 21–27 July 2000, published in honour of Liliane Dulac* (3 Vols, Glasgow, 2002); Earl Jeffrey Richards, ed., with Joan Williamson, Nadia Margolis and Christine Reno *Reinterpreting Christine de Pizan* (Athens, Ga, and London, 1992); Margaret Zimmermann and Dina De Rentiis, eds, *The City of Scholars: new approaches to Christine de Pizan* (Berlin, 1994); L. Dulac and B. Ribémont, eds, *Une femme de lettres au Moyen Age: études autour de Christine de Pizan* (Orléans, 1995); Marilynn Desmond, ed., *Christine de Pizan and the Categories of Difference* (Minneapolis, Minn., 1998).

54. Earl Jeffrey Richards, 'Christine de Pizan and Jean Gerson: an intellectual friendship' in John Campbell and Nadia Margolis, eds, *Christine de Pizan 2000: studies on Christine De Pizan in honour of Angus J. Kennedy* (Amsterdam and Atlanta, Ga, 2000), pp. 197–208; Nathalie Nabert, 'Christine de Pizan, Jean Gerson et le gouvernement des âmes', in *Au champ des escriptures*, pp. 251–68; Mary Agnes Edsall, 'Like Wise Master Builders: Jean Gerson's ecclesiology, *Lectio Divina* and Christine de Pizan's *Livre de la Cité des Dames*', *Medievalia et Humanistica* 27 (2000), 33–56.

55. On this process, see, for example, Kathryn Kerby-Fulton, '*Eciam Lollardi*: some further thoughts on Fiona Somerset's "*Eciam Mulier*: Women in Lollardy and the problem of sources"' in Linda Olson and Kathryn Kerby-Fulton, eds, *Voices in Dialogue. Reading women in the Middle Ages* (Notre Dame, Ind., 2005), pp. 261–78. Kerby-Fulton's essay ranges more widely than its title suggests, situating the specific problem of 'women in Lollardy' with the broader context of the impact of women on late-medieval European religious and theological developments.

56. Alexander Murray, *Reason and Society in the Middle Ages* (Oxford, 1978), pp. 6, 401, 404.

57. Boureau, 'Intellectuals in the Middle Ages', p. 148.

58. Gabrielle Spiegel, *The Past as Text. The theory and practice of medieval historiography* (Baltimore, Md, and London, 1997), p. 25.

6

intellectual history and the history of political thought

richard whatmore

The historian, investigating any event in the past, makes a distinction between what may be called the outside and the inside of an event. By the outside of the event I mean everything belonging to it which can be described in terms of bodies and their movements: the passage of Caesar, accompanied by certain men, across a river called the Rubicon at one date, or the spilling of his blood on the floor of the senate-house at another. By the inside of the event I mean that in it which can only be described in terms of thought: Caesar's defiance of Republican law, or the clash of constitutional policy between himself and his assassins. The historian is never concerned with either of these to the exclusion of the other. He is investigating not mere events (where by the mere event I mean one which has only an outside and no inside) but actions, and an action is the unity of the outside and inside of an event. He is interested in the crossing of the Rubicon only in its relation to Republican law, and in the spilling of Caesar's blood only in its relation to a constitutional conflict. His work may begin by discovering the outside of an event, but it can never end there; he must always remember that the event was an action, and that his main task is to think himself into this action, to discern the thought of its agent.

R. G. Collingwood, *The Idea of History* (1946)[1]

R. G. Collingwood's description of the difference between the 'inside' and 'outside' of an event had profound implications for historians of political thought in the 1960s, when it played a role in inspiring the articulation of the approach to intellectual history that has come to be known as that of 'the Cambridge School'.[2] Collingwood's choice of the example of Ceasar's death at the hands of assassins seeking to save the republic was fortuitous, in so far as the work of those associated with the Cambridge School has heavily contributed to a remarkable upsurge of interest in republicanism as an historical tradition of political argument.[3] Much has been written about this development since the publication of John Pocock's *The Machiavellian Moment* in 1975 and Quentin Skinner's *The Foundations of Modern Political Thought* in 1978; with the recent appearance of reassessments of historical republicanism by these authors, a re-evaluation of the subject is timely.

i

The scholarly activities that help to explain contemporary interest in republicanism are varied, and often characterised by the role of unintended consequences. Two important moments in the story occurred in 1949 and in 1960, when the Cambridge historian Peter Laslett published an edition of the writings of Sir Robert Filmer, followed by a critical edition of John Locke's *Two Treatises of Government*.[4] Laslett revealed that Filmer's *Patriarcha* had been composed prior to his other writings, and published posthumously between 1679 and 1680. More surprisingly, Locke's text, rather than being a defence of the 'Glorious Revolution', as it had traditionally been described, having appeared in print from 1690, was shown to have been written around 1681, at a time when Whigs of Locke's stamp were contemplating violence against the Stuart court. Laslett's editions raised questions for historians concerning the relationship between an author's intentions in writing a text and the intentions behind its publication, causing in the process a wholesale re-evaluation of late seventeenth-century political thought.

Between Laslett's editions, three works were published which revised the way historians thought about early modern intellectual life, and complicated inherited perspectives on the origins and elements of 'modern' political thought. In 1955, Hans Baron's *The Crisis of the Early Italian Renaissance* described the reassertion of the classical republican *vita activa* by Colluccio Salutati and Leonardo Bruni, the great Quattrocento Chancellors of the Florentine Republic, in the face of the threat of monarchical tyranny represented by the Visconti of Milan.[5] In 1957, John Pocock published *The Ancient Constitution and the Feudal Law*, which further questioned accepted understanding of seventeenth-century thought, by revealing the significance to contemporaries of a 'language of politics' beyond jurisprudence and governance, tying together disputes about the antiquity of common law and parliament and the meaning of the Norman Conquest. In 1959, Caroline Robbins published *The Eighteenth-Century Commonwealthman*, which traced the development of Old or Real Whig arguments, in England, Scotland, Ireland and North America, for the rotation of public offices, for the relevance of natural rights to political liberty, and for the necessity of a separation of constitutional powers in free states, between the Restoration and the assertion of North American independence.[6] Revisionist studies emphasising the importance of contextual readings in the history of political thought appeared throughout the 1960s, with leading instances having been produced by Felix Gilbert, Bernard Bailyn, Gordon Wood, and John Dunn.[7] Linking together such writings has become commonplace, as they challenged the position of Locke at the centre of a putative 'liberal Enlightenment', and focused on the reconstruction of neglected political languages, and especially republicanism as an identifiable tradition of thought.[8]

The methodological justifications of this historical turn, which appeared concurrently with the substantive reinterpretations of early modern history, are equally well known. At least three examples, by John Pocock, John Dunn and Quentin Skinner, all associated with the University of Cambridge, have come to be seen as classic statements about the practice of intellectual history in the area of political thought.[9] Skinner's 'Meaning and Understanding in the History of Ideas' proved especially influential, and has been described as the manifesto of the 'Cambridge School', in part because it identified opponents most clearly, and condemned with particular verve descriptions of classic texts as the 'sole object of enquiry', and as contributing to a 'dateless wisdom'. Skinner's intentions in writing 'Meaning and Understanding' have been summarised many times, but it would be difficult to better Pocock's own recent synopsis:

[Skinner] demonstrated that much of the received history of [political thought] suffered from a radical confusion between systematic theory (or 'philosophy') and history. The greater and lesser texts of the past were interpreted as attempts to formulate bodies of theory whose content had been determined in advance by extrahistorical understandings of what 'political theory' and 'history' should be and were. This confusion led to errors including anachronism (the attribution to a past author of concepts that could not have been available to him) and prolepsis (treating him as anticipating the formation of arguments in whose subsequent formation the role of his text, if any, had yet to be historically demonstrated). After treating these fallacies with much well-deserved ridicule, Skinner contended that the publication of a text and the utterance of its argument must be treated as an act performed in history, and specifically in the context of some ongoing discourse. It was necessary, Skinner said, to know what the author 'was doing': what he had intended to do (had meant) and what he had succeeded in doing (had meant to others). The act and its effect had been performed in a historical context, supplied in the first place by the language of discourse in which the author had written and been read; though the speech act might innovate within and upon that language and change it, the language would set limits to what the author might say, might intend to say and be understood to say. The language would further have been a means by which the author acquired and processed information about the historical, political, and even material situation in which he lived and was acting; and though a great deal of 'political thought' had been second-order language – thought about the language in which politics was thought about – it was possible to enlarge the 'context' (from now on a key term for

Skinner and his readers) beyond the language to its referents (though once the historian began using referents not fully articulated in the language, the danger of prolepsis, and perhaps its necessity, must return).[10]

In each of their essays, Pocock, Dunn and Skinner, proposed that texts be treated as products of specific historical contexts. In working out the meaning of the texts, Dunn and Skinner identified the intentions of an author as the principal guide to the nature of the text, although neither unproblematic as an intellectual objective, nor sufficient in terms of understanding an author's work.[11] The three authors professed themselves to be sceptical of approaches founded on the presumption of fixed concepts of historical analysis, of metatheoretical assumptions about human nature, and of opaque or ahistorical theoretical vocabularies. Within the historical fraternity, one goal of the attempt to analyse utterances in political texts as 'speech acts' was to liken them to the 'acts' analysed by fellow historians, thereby countering prevalent Namierite opposition in Anglophone circles to the study of ideas. This aspect of the enterprise has been the most obviously successful, as a generation of scholars has confidently described themselves as intellectual historians, formed professional associations, and acquired academic positions in this field.[12] Defence of the contextualist method attracted philosophical evaluation. In consequence, the methodology of the Cambridge School has been attacked by Straussians, Marxists, empirical political scientists (behaviourists) and post-structuralists. It has become a subject of study in itself, and recent commentators have focused on the relationship between the Cambridge School, German hermeneutics and French social theory, in addition to English-language political theory and political and analytical philosophy.[13] Such developments reflect the fact that the attempt to contextualise political thought raised fundamental questions about the division of academic labour between university faculties in the late twentieth century, and more especially the relationship between political thought and political philosophy.[14]

ii

Among the major questions for historians in the 1970s was the extent to which use of the 'Cambridge' method would continue recent contextualist work in altering contemporary perspectives upon the history of political thought? Among a range of revisionist interpretations of canonical figures in the history of political thought, Pocock's *Machiavellian Moment* and Skinner's *Foundations* provided notable, some would say seminal, answers. Each work generated a great deal of scholarly comment, much of which has linked them together as the most important studies to date of early modern republicanism in Europe and the Atlantic World.[15] This was especially the case after the publication of Skinner's work on Machiavelli and the republican idea of political liberty between 1981 and 1984.[16] *The Machiavellian Moment* and *The*

Foundations were of course concerned with far more than the reconstruction of a neglected republican tradition in politics. Pocock's broader concern with the understanding of time in history went far beyond historical republicanism, howsoever defined. The second volume of Skinner's *Foundations*, *The Age of Reformation*, describing the gradual identification of the state with an artificial person standing above the unruly elements of political society, being a potential solution to endemic religious conflict, was not concerned with republicanism at all. Nevertheless, Skinner and Pocock agreed on the importance in early modern history of the articulation of a view of human flourishing developed in the Italian city republics from the thirteenth century onwards, in which a life lived in an independent republic was praised as natural to man, and in which civic activity to sustain this state was described as being the highest secular aspiration.[17] Pocock's recent restatement of what he meant by 'the Machiavellian Moment' is very clear on this point:

> It can denote the historic 'moment' at which Machiavelli appeared and impinged upon thinking about politics, and either of the two ideal 'moments' indicated by his writings: the moment at which the formation or foundation of a 'republic' appears possible or the moment at which its formation is seen to be precarious and entails a crisis in the history to which it belongs. I see these moments as inseparable, and there thus arises 'the Machiavellian moment' as that in which the republic is involved in historical tensions or contradictions which it either generates or encounters. I go on to present much, but not all, of early modern political thought as the experience and articulation of this 'moment'.[18]

While Skinner's *Foundations* ended with the resistance theories of the early seventeenth century, Pocock's history of civic humanism extended chronologically and geographically, describing Harrington's adaptation in the mid-seventeenth century of Machiavelli's perspective on the history of Rome, according to which the practice of arms and inheritable landed property were the precondition for the enjoyment of liberty and the exercise of civic virtue. Pocock went on to detail the controversy over this perspective in the eighteenth century, in the very different context of ideas about liberty and autonomy that accompanied the end of the wars of religion and the rise of large and competing commercial monarchies in Europe.[19] Maintaining these new forms of state required the development of standing armies and public credit; these were defended by association with modern forms of politeness, and praised by such men as Daniel Defoe alongside consumption and financial independence, in societies organised in accordance with the division of labour.[20] 'Neo-Harringtonians', as Pocock termed them, such as Andrew Fletcher, favoured ancient virtue, protected and sustained by militias and an elite of landowners, whose interest in the state ensured their political wisdom and the moderation of their laws. The neo-Harringtonians despised modern politeness, believing it to entail the corrosion of masculinity, the growth of forms of corruption accompanying the rise of parties and the specialist politician, and far greater uncertainty in civil society and in politics, exemplified by the reliance of the state upon the expertise of the stock-jobber.

The clash between 'ancient' and 'modern' ideas about liberty has been characterised by Pocock as a debate in which the ancients saw liberty as a 'direct act of the personality ... the liberty of the hedgehog who knows himself but may know nothing else' and in which the moderns supposed liberty to entail the mediation of the personality, 'through all the multifarious activities which may relate humans to one another in society ... the liberty of the fox, who knows so many things that he may have no self left to know'.[21] In the Atlantic world the controversy was characterised by concerns about the likely stability of real and of movable property, and was in turn complicated both by the domestically uncertain identity of composite states comprised of different 'nations', and by discussions in Protestant and Catholic states of the likely consequences of commercial society for religious belief, some of which questioned the compatibility of different forms of Christian worship with a polity competing for survival by commerce and by war.[22] This state of affairs in turn generated an influential literature of jeremiad, which was particularly marked in the writings of historians, politicians and philosophers in the late eighteenth century:

The individual who possesses the means of his own freedom is liable to regress into barbarism, unsupported by the freedom of others; the individual whose freedom consists in the exercise of diverse capacities, but who never brings them together in the performance of public acts in his own person, is liable to progress into corruption and find himself subject to tyranny. There is no ideal moment in history, and though we may imagine a species of freedom which consists in the freedom to move prudently between the ancient and the modern poles, the exercise of such freedom depends upon the maintenance of the unity of personality necessary to act in history, and history has become a process of rendering that unity precarious. There is consequently a species of historicism, in which many or most thinkers of the eighteenth century can be said to have been involved.[23]

Pocock has always contended that such issues remain identifiable in the political cultures of modern states, and particularly the North American Republic, and this has been one of the reasons for the controversial reception of his work within and beyond the academy.[24] Amongst historians, the astonishing impact of the book has largely been due to the ease with which Pocock's recovered republican language could be used to make sense of authors' writings at times and in states not considered in *The Machiavellian Moment*. This was particularly the case with the writings of those who called themselves republicans. A minor, but not insignificant, example can be drawn from the letters of the Genevan pastor Etienne Dumont, writing in Paris in April 1789 before the opening of the Estates General. Dumont was observing the private meetings held 'to consider what steps the commons ought to adopt', prior to the gathering of the primary assemblies of the districts of Paris:

> I was present at two of these meetings, and I was astonished to observe what republican ideas prevailed in them, and how much the Parisians are weaned from their ancient love of monarchy. They seem to consider the principles of Rousseau's *Social Contract* as the only sound principles of government.[25]

Later in the same year, on August 15, Dumont considered manifestations of a transformation of national character that he believed had occurred in France over a singularly short period of time, and identified the root cause:

> The people, universally, regard their representatives as their protectors and deliverers; all affection for the king appears to have diminished, in proportion as that for the [national] assembly has increased. If a Machiavellian republican had sought to degrade and ruin the court by his insidious counsels, he could not have suggested any more effectual measures than those it has pursued.[26]

Dumont believed that the commercialisation of advanced European societies was inevitable, but was also sympathetic to the neo-Harringtonian critique of the excesses of Britain's mercantile system. He believed in the progress of moderate Socinianism as a means of addressing the worst effects of luxury and inequality, and while remaining a convinced republican with respect to Europe's smaller states, he was horrified by the prospect and growth of Machiavellian republicanism in larger states and established monarchies like France. Dumont considered himself to be an ardent republican but coupled this with vehement opposition to republicanism in France. He persuaded the publisher Le Jay to burn an edition of Milton's anti-royalist writings, and refused to translate Thomas Paine's journal *The Republican* into French. Such views, once deemed inexplicable or unworthy of scholarly analysis, acquire a singular clarity in the light of Pocock's description of the clash between ancient and modern values.

Among the major historiographical developments since the appearance of the *Foundations* has been Skinner's shift of interest beyond the early seventeenth century, and towards the history of ideas about liberty from Hobbes to the present.[27] This has accompanied alterations to his understanding of the nature of republican liberty in early modern history. In the *Foundations*, Skinner narrated a distinctive story, which challenged Pocock's account by emphasising the differences between the Greek and the Roman political heritage, arguing specifically that Pocock had overestimated the importance of Aristotle's *Politics* as a source of Renaissance understanding of what it meant for citizens to live as equally sovereign agents in a law-governed republic. For Skinner, civic humanism was more heavily indebted to Cicero, who had extolled the life of civic virtue and republican citizenship that was possible under a free government.[28] From Pocock's perspective, the fundamental difference between his view of civic humanism and Skinner's lies in the importance of justice as a political value. While Cicero encouraged the enjoyment of all the goods of human society and their just distribution, entailing the development of a moral code capable of defining and sustaining the life fully lived, Machiavelli was less concerned with justice and morals than with *virtù*, and placed personal autonomy and public participation at the centre of his political philosophy.[29] For Pocock, historical Ciceronianism was more a moral philosophy than a doctrine for political action, and it was Machiavelli's distinctive vision which defined republican thought and practice in Europe's small states, enjoying another history when it was adapted to the circumstances of Europe's larger monarchies and their colonial empires. Once more, the recent 'Afterword' to *The Machiavellian Moment* is revealing of Pocock's response to Skinner's criticism, especially when set beside a similar passage from the third volume of *Barbarism and Religion*. Together they make the case for placing Skinner's approach within the history of jurisprudence, rather than being directly related to the republicanism of Machiavelli:

Political thought, theory, or philosophy – pervaded as they have been at every point by jurisprudence – are, we might simplify the account by saying, the ideology of liberal empire; what has come down to us from the republic is another matter. Historiography – meaning here the construction of grand historical narratives – has taken two distinguishable courses: the one recounting the transformation of republic into empire, the other maintaining that *libertas* and *imperium* are both inseparable and mutually destructive. In the work I am doing while I write this afterword, on the history of the topos of Decline and Fall, I have been led to conclude that – though Cicero himself was a martyr to the republic in whose downfall he perished – the 'Ciceronian' ideal of citizenship discovered by Skinner in the thirteenth century was by no means incompatible with the proposition that the civic virtues might be practiced under the rule of law and a just prince, so that Augustus, Trajan, or Justinian ruled over men free in the sense that there was law to hand to which they could appeal. Hans Baron's Florentines, on the other hand, two centuries later, were capable of insisting that under the Caesars *libertas* disappeared, with the result that the princes became tyrants and monsters, and the citizens had no longer the virtues necessary to maintain the empire it had acquired against the barbarians.[30]

In *The Foundations of Modern Political Thought* Quentin Skinner laid emphasis on the praise of the civic life to be found in Brunetto, Ptolemy, and other writers of the thirteenth – and fourteenth-century Italy including Florence. He saw this as Ciceronian even more than Aristotelian in its intellectual foundations, and as massively anticipating the 'civic humanism' of Baron's Renaissance thinkers. It may be noted that his work of the above title opened with an account of Otto of Freising's attempt to explain Italian civic liberties to his readers in imperial Germany, as if the 'foundations of the modern' had been laid by forces akin to Baron's but operating centuries earlier. In the present work, which has pursued the history of the *translatio imperii* and is about to pursue its replacement by 'decline and fall' – but is not so far committed to the view that this entails a replacement of 'medieval' by 'modern' – it has been argued that the 'republicanism' of the *trecento* was perfectly compatible with the mythos of Eusebian sacred monarchy and the *translatio imperii*. If it can be established that an abandonment of translation for declination occurred within the same narrative as that recounted by Baron, we shall have evidence that a breach of some kind occurred and can proceed to enquire whether it was connected with a new understanding of republican liberty and citizenship – a return, perhaps, from *translatio imperii* to *libertas et imperium*. It would be paradoxical if this re-assertion of values so ancient as to be pre-Christian proved to be the foundation of the 'modern'.[31]

In addition to the substantive reinterpretation of many of the most significant events of early modern history, Pocock's *Machiavellian Moment* followed *The Ancient Constitution and the Feudal Law* in leading historians to broaden their definition of early modern political theory, which was shown to encompass civil law, theology and political economy, by contrast with many narrower twentieth-century conceptions, which had inspired altogether different readings of history.[32] Yet Pocock's own description of the intellectual journey he has undertaken since *The Machiavellian Moment* contrasts the insufficient breadth of his notion of the political employed in that book with the more theologically embedded perspective on politics which has characterised his subsequent historical writing, and most directly that concerning Edward Gibbon's *Decline and Fall*.[33] Although Pocock continues to defend the central thesis of *The Machiavellian Moment*, it is certain that he no longer sees the Machiavellian tradition of republican writing as encompassing so much of early modern European political thought as he did in 1975. The level of contextual complexity in his subsequent writings surpasses that of other intellectual historians, and numerous reviewers have commented on the multiple, interconnected and mutually-influencing languages that now make up his portrayal of the eighteenth-century intellectual landscape.[34] While a summary account of these 'new' languages is beyond the scope of this chapter, from the perspective of historical republicanism Pocock is now extremely careful about his use of the terms 'republic', 'republicanism' and

'Atlantic republican tradition'. It is likely that he would be equally cautious in using the definite article before 'Machiavellian Moment'.

iii

In Skinner's revised perspective, Pocock's reconstruction of civic humanist politics underestimated the influence of Machiavelli's statement in the *Discorsi* that a community could be self-governing whether it was a republic in name or a principate. In consequence, it becomes a mistake to call the tradition of political reflection associated with Machiavelli essentially republican, and this in part explains the manifest interest in civic humanism across Europe's monarchies. The history of seventeenth-century England provides a case study. While concurring with the view that the classical legacy became significant in politics from the time when humanist values began to shape a particular culture, thereby defining an English 'renaissance', for Skinner the key period in the formation of a distinctive political theory occurred after the regicide of 1649, when John Milton, Marchamont Nedham and others were commissioned to defend the new state.[35]

Milton and Nedham were convinced that the actions of a political body had to be determined by the will of its members. They advocated representation rather than democracy, defining democracy as a system in which the people acted as their own government.[36] Representation, by contrast, was one means of promoting government by the most virtuous and most wise, although Milton and Nedham recognised that it would also be necessary to have laws that encouraged *virtù* in the people, raising the possibility of laws 'to coerce the people out of their natural but self-defeating tendency to undermine the conditions necessary for sustaining their own liberty'.[37] Arguing against the exercise of prerogative power by governments, Milton and Nedham held that the existence of such executive powers rendered subjects or citizens unfree, because of the threat to life and to property inherent in the existence of such powers. As such, they countered Hobbes's definition of a free man as someone 'not hindered to do what he has a will to' by praising civil liberty and political action with reference to Sallust, Seneca and Tacitus. In consequence, Skinner has termed the doctrine of Milton and Nedham 'the neo-roman theory of free states and free citizens'. The essence of the theory was nicely captured in Harrington's response to Hobbes's famous claim that the liberty proclaimed on the turrets of the city of Lucca was illusory, because the extent of the citizens' freedom within the law was no greater than that enjoyed under the Sultan. Harrington stated that 'even the greatest bashaw in Constantinople is merely a tenant of his head', enabling Skinner to conclude:

> When the neo-roman theorists discuss the meaning of civil liberty, they generally make it clear that they are thinking of the concept in a strictly political sense. They are innocent of the modern notion of civil society as

a moral space between rulers and ruled, and have little to say about the dimensions of freedom and oppression inherent in such institutions as the family or the labour market. They concern themselves almost exclusively with the relationship between the freedom of subjects and the powers of the state. For them the central question is always about the nature of the conditions that need to be fulfilled if the contrasting requirements of civil liberty and political obligation are to be met as harmoniously as possible. When considering this question, these writers generally assume that the freedom or liberty they are describing can be equated with – or, more precisely, spelled out as – the unconstrained enjoyment of a number of specifically civil rights ... the theory of liberty they espouse appears to me to constitute the core of what is distinctive about their thought. More than their sometimes ambiguous republicanism, more even than their undoubted commitment to a politics of virtue, their analysis of civil liberty makes them out as the protagonists of a particular ideology, even as the member of a single school of thought.[38]

Using the structure of arguments about liberty to distinguish between schools of political thought from the time of the English Revolution has led Skinner to identify two antagonistic conceptions of political freedom. Early modern history, as he describes it, is largely an argument between exponents of the 'roman' view of liberty, describing the life of the active citizen who makes the law and seeks to live in accordance with virtue, and the 'Gothic' view that liberty is an attribute of an owner of land whose rights are protected by law, who did not make the law, but who seeks to live with minimal interference under the law. In the seventeenth century, the major exponent of the latter view was Hobbes, and neo-romans, such as Nedham, Milton, Harrington, Algernon Sidney and Henry Neville, sought to combat Hobbes, without falling into the 'monarchomach' camp of Henry Parker, who argued that the sovereign people delegated their liberty to their rulers.[39] Skinner has stated that neo-roman theory was 'the most formative influence' on those who 'challenged the government of Charles I and instituted the first and only British republic'. In the eighteenth century he has claimed that neo-roman theory proved hugely influential in North America and in revolutionary France.[40]

Skinner considers the neo-roman view to have declined for two reasons. Firstly, 'with the extension of the manners of the court to the bourgeoisie in the early eighteenth century, the virtues of the independent country gentleman began to look irrelevant and even inimical to a polite and commercial age'. Secondly, and more importantly, liberal critics, from William Paley to Jeremy Bentham and later Henry Sidgwick, accused the neo-roman writers of being confused about liberty. Skinner identifies as neo-romans Viscount Bolingbroke, Richard Price and Thomas Jefferson in the eighteenth century, and in the nineteenth century finds examples of neo-roman argument in Chartist demands for annual parliaments and equal electoral districts, and in John Stuart Mill's attack upon the subjection of women.[41] While the neo-romans asserted that 'to live in a condition of dependence is in itself a source and a form of constraint', for liberals the potential of being coerced was irrelevant up to the point at which free citizens *were* in fact coerced.[42] Skinner's claim is that the neo-roman theory was not a branch of liberalism. He has further stated that it is important to reconstruct the story of its rise and decline, because it provides a perspective on the history of liberalism which might lead us to question our presumption of its supremacy among potential political doctrines:

> The intellectual historian can help us to appreciate how far the values embodied in our present way of life, and our present ways of thinking about those values, reflect a series of choices made at different times between different possible worlds. This awareness can help to liberate us from the grip of any one hegemonal account of those values and how they should be interpreted and understood.[43]

Skinner's historical research thereby neatly dovetails with his separate commitment to a political philosophy defining an individual's freedom from domination in a broader sense than classical liberalism, and justifying more directly the claim that to be unprotected by law or by the mechanisms of parliamentary representation and consent is to be in the condition of a slave.

Skinner's restated 'foundations of modern political thought' have been criticised on numerous grounds.[44] The concern with a fixed 'concept' of liberty defining a developing historical school of thought identical with Skinner's personal political philosophy, and deemed to be of direct relevance to modern problems in politics, has been called Whiggish, and has been likened to the approaches founded on prolepsis that Skinner attacked in 'Meaning and Understanding'. The fact that none of Skinner's cast called themselves neo-romans, or identified themselves as a separate group with the other authors listed as neo-romans, has led to the accusation that Skinner's categories are ahistorical, or at best misleading, because they were not used by the historical actors concerned (and what an author calls himself or herself, and the labels contemporaries gave to them, have always been seen by advocates of contextualism to be a crucial guide to an author's intentions and the understanding of the reception of a text). Despite being the person who has done the most in recent years to bring philosophers and intellectual historians into dialogue, Skinner has been accused of being too much of an historian and too much of a philosopher by members of each tribe. His definition of politics as comprising precise but narrow concepts, explicable by means of a critique of Berlin's discussion of liberty in the mid-twentieth century, has also been described as too blunt an instrument of early modern historical analysis, in part because of Pocock's success, and that of Jonathan Clark, in studying early modern politics together with theology and political economy.[45] Skinner's claim that the neo-romans' idea of 'civil liberty' was 'strictly political' is odd in that it implies that the neo-romans were uninterested in the nature of the Christian polity, and in the forms of international competition considered vital to maintain security. To give one example, it is remarkable in the light of recent historiography to imply that as a neo-roman Richard Price's notion of civil liberty was independent of his theology, and more especially independent of his work on the operation of markets. A further point of dispute arises once the sheer variety of republican (and by extension neo-roman) thought is acknowledged, following as it does the various meanings of the term *res publica*. In failing to make clear the difference between republican and neo-roman argument, Skinner has been accused of vacillating over his commitment to the historiographical assumption that defence of the *res publica* was a coherent and identifiable tradition: an assumption that has been described as no longer tenable.

Some of these criticisms can be seen to have missed their target once it is accepted that Skinner's main aim has been to make sense of a complicated controversy over the meaning of liberty in seventeenth-century England, while valuably reminding his audience that liberalism is itself an historical construct. Accepting that Skinner remains an intellectual historian, attempting to do what he has done so successfully in the past, is the characteristically generous verdict of Pocock:

> I have not made use of Skinner's 1978 generalization that 'modern' political thought is concerned with the increasing impersonality of the state – is he still operating within that paradigm? If so, he will not arrive at conclusions at variance with mine, but he may narrate different histories, coexisting and interacting with those I have been relating. We are in a history where, even within the same texts, many things not necessarily compatible may be found going on concurrently. There need hardly be a master narrative of modern history that excludes or absorbs all other interpretations.

If the rise of the impersonal state remains central to Skinner's interpretation of modern political thought, a genuine question is the nature of the neo-roman response, from the late seventeenth century, to the great change in attitudes that accompanied the rise of commercial monarchies that were able to use public credit to fight wars.[46] This difficulty was identified some time ago by John Dunn, at a time when Skinner was still seeking to define a republican tradition across Europe, and was identifying it with the history of particular republican states:

> It was certainly still a very open question, as the nineteenth century dawned, how far a modern constitutional republic could indeed hope to provide

a compelling recipe for political legitimacy of any real durability. That question has now been answered, at least for the time being. The question which remains is just what it was about this somewhat hazy formula which enabled it to provide such a recipe. One answer might be that it was its residual republicanism. A second might be that it was its comparatively novel (and as yet somewhat fair-weather) constitutionalism. But a third, and in the end a strategically more compelling, answer will have to be that it was its modernity: not its detailed constructive statecraft (which was often as accident-prone as anyone else's) but its relatively steady imaginative acceptance of the economic limits to modern politics, and its continuing readiness to adjust to these limits in the face of disappointing experience. This is the judgement which I shall try to develop ... I wish to contrast it in its entirety with the model of classical republicanism as Quentin Skinner has anatomized this. The city republics of early modern Italy, drawing on Sallust and Cicero, but also drawing more directly on their own political experience, praised their own political form for its direct contribution to the attainment of *grandezza* and for securing a free way of life. In Machiavelli's arresting reworking of their vision, the free way of life and the degree of popular commitment to a common good for which that way of life was in his view a clear necessary condition were both in the end vindicated in the last instance because of their indispensable joint contribution to expanding the power and wealth of the community. But, by the time that Montesquieu came to write, it was not necessary to emulate his sceptical sophistication to see that the classical republicans were palpably wrong, at least on this score ... What we can safely retain from these severely preliminary considerations is the simple judgement that it was always likely to prove true, and has in the end proved true, that the key dimension of practical comparison for modern models of political legitimacy is in their capacity to furnish their political clients with security. The provision of security to human populations is an inherently complicated and accident-prone endeavour; and judgements about how it is best undertaken have naturally shifted extensively over the last few centuries.[47]

Dunn was here questioning the relevance of Skinner's republicanism to contemporary politics, but a related critique expressed by eighteenth-century authors evaluated republican political messages for modern states.[48] David Hume's essay 'Of Civil Liberty' provides one of the best-known illustrations, in stating that 'Trade was never esteemed an affair of state till the last century, and there is scarcely any ancient writer on politics, who has made mention of it.'[49] Hume continued to defend the position that 'the extensive commerce' so important to moderns would thrive in free governments, and in defining free government he believed the history of republics from ancient times to his own was always instructive. At the same time, he was certain that political thought had been transformed not only by the development of commerce as a form of reason of state, but also by the struggle for international supremacy between France and Britain. The former state provided examples for Hume's contention that 'monarchical government seems to have made the greatest advances towards perfection', raising the possibility that once ministers in Europe's courts addressed iniquities in taxation, 'the differences betwixt their absolute and our free [government], would not appear so considerable as at present'.[50] Of equal significance were the changes to the understanding of monarchy that accompanied the success of mixed government in Britain after 1688.[51] Free states such as Britain had to address the great 'source of degeneracy', consisting in 'the practice of contracting debt'. Hume concluded that such issues surrounding the security of modern states had become a major fault-line that led to the formation of new schools of thought in politics from the late seventeenth century onwards. This was recognised in Pocock's distinction between 'ancients' and 'moderns', but cannot be discerned in Skinner's description of neo-roman argument.[52]

It is also unlikely that theories formulated in defence of a regicide state could have played the role that is imputed to them by Skinner during the next 150 years of British history. All English radicals before the French Revolution accepted that the post-1688 monarchy, however threatened liberty might have been by the perceived growing powers of the executive, was maintaining Britain by securing the state domestically and internationally. This was one of the reasons why Richard Price denied so forcefully the imputation that Protestant Dissenters such as himself were republicans.[53] G. E. Aylmer once questioned whether Skinner's *Foundations* overemphasised historical continuity, and underplayed 'discontinuities, resulting either from individual genius or from contingent circumstances'.[54] A similar point had earlier been made about historians of republicanism by Franco Venturi, when he argued that attempts to generalise about the classical legacy in early modern thought were likely to be misguided because they understated the differences separating ancient and modern political realities.[55] Venturi followed Hume in accepting that the republican heritage had been transformed by the prospect of commercial monarchies with the military force capable of destroying small states with relative ease. To many mid-eighteenth-century commentators, this made republics an outdated political form. Events in North America and in France ensured that republicanism became once more a significant political idiom. For Venturi, the sudden republican renaissance required explanation, and famously he contended that the republican patriotism characteristic of the small surviving republics across Europe began to influence the reform projects of the monarchies which found themselves facing bankruptcy.

If the writings of some of Venturi's cast of eighteenth-century republicans are examined, a significant fact emerges that many citizens living in the surviving city republics expressed exactly the perspective on the nature of liberty that Skinner has identified as neo-roman. Examples can be found at Geneva, where opponents of the existing magistrates were demanding a new law-code that would outlaw the exercise of prerogative powers. The group pushing for reform called themselves the *'représentants'*, because they continually exercised their right to carry or 'represent' their grievances to the magistrates. The claim uniting the *représentants* was that since the beginning of the century there had existed at Geneva a system calculated for subduing the citizens, and forcing them to silence by authority and fear.[56] They attacked the magistrates for becoming an aristocracy seeking 'to awe the people by reserve and stateliness, and entirely forgot that the only possible method to govern a republic, is to acquire a dominion over the hearts of its citizens'.[57] In the 1760s the group comprised lawyers such as François D'Ivernois, Jean-Louis Delolme and Jacques-Pierre Duroveray, merchants such as Etienne Clavière, and natural philosophers such as Jean-André Deluc. In the summary of their political beliefs published in the early 1780s, these *représentants* identified the source of their politics as 'the excellent works which appeared [at the beginning of the century], on the science of government'; a science which 'till then existed only in maxims scattered through the works of Tacitus, and some authors far beyond the reach of the vulgar'.[58]

The first contrast between *représentant* opinion and that of the neo-romans described by Skinner lies in their political economy. For the *représentants*, being 'strictly political' was inconceivable, as they directly linked the reform of government with the new kinds of industry that were transforming Genevan commerce. That industry was the fruit of liberty was a fundamental tenet. They opposed mercantile controls upon trade and had faith that a moralised form of commercial society was a possibility where the political and economic laws of the state countered aristocracy.[59] Constitutionally this meant annual elections for magistrates, the right to remove corrupt magistrates, and a charter of individual rights that protected against the encroachment of the rich. Their views of the relationship between political and economic justice were indebted to the forms of Calvinism focused on the practice of morality that had developed at Geneva under the influence of J.-A. Turettini, Werenfels, Osterwald, Jacob Vernet and others.[60]

All of the *représentants* supported the 'revolution' of 1782, which initially saw the existing magistrates expelled from the city, before they were returned through a French-led invasion. The now-exiled *représentants* then acknowledged that the success of their movement depended upon gaining the support of larger powers, and especially Britain or France. By the mid-1780s, D'Ivernois, Deluc and Delolme had settled in Britain. Deluc gained fame as a natural philosopher and influence as Reader to Queen Charlotte. Delolme had already come to prominence as the author of the *Constitution d'Angleterre* (originally published in 1771). Duroveray and Clavière came to live at Paris, and were the authors of many of Mirabeau's speeches in the National Assembly in the first years of the Revolution. Once these 'neo-romans' considered the reform of larger states their political ideas underwent a profound and unpredictable transformation. Clavière, who became a minister of state in the Girondin ministries under Roland, became a convert to republican empire and sought the means of increasing popular patriotism in the hope of making better citizens, soldiers and traders. By contrast, his erstwhile friend D'Ivernois opposed the revolutionary experiment with the union of democracy and aggressive patriotism. In England he became a leading critic of the French Republic, and always argued that republicanism could never in the modern world be envisaged beyond the walls of a small and pacific city.[61] Duroveray shared this opinion, advocating French adoption of the British model of mixed monarchy and constitutional government.

General lessons cannot be derived from such particular historical experiences, but they at least illustrate the mutability of individual political commitment and republican language. They also underline the necessity of linking politics with political economy and with theology in early modern intellectual history, which in turn problematises definitions of opposed historical 'traditions' of thought spanning centuries, nations and cultures. Advances in our knowledge of historical context have made our sense of the meaning of 'the republic' and 'republicanism' more nuanced and complicated. A pluralist approach to early modern republicanism, once favoured by Venturi, now characterises Pocock's and Skinner's works. With the politics of virtue always in the vanguard, republican ideas are increasingly related only to particular historical moments.[62]

notes

1. R. G. Collingwood, *The Idea of History* (Oxford, 1946), p. 213; see also *An Autobiography* (Oxford, 1939), pp. 81–8.
2. Q. Skinner, 'Meaning and Understanding in the History of Ideas', 'Social Meaning and the Explanation of Social Action', 'Some Problems in the Analysis of Political Thought and Action', 'A Reply to My Critics', in J. Tully, ed., *Meaning and Context: Quentin Skinner and his Critics* (Princeton, NJ, 1988), pp. 55–6, 65, 80, 103, 233–4; *Liberty before Liberalism* (Cambridge, 1998), p. 102; 'The Rise of, Challenge to and Prospects for a Collingwoodian Approach to the History of Political Thought', in D. Castiglione and I. Hampsher-Monk, eds, *The History of Political Thought in National Context* (Cambridge, 2001), pp. 175–88.
3. A. Patten, 'The Republican Critique of Liberalism', *British Journal of Political Science* 26 (1996), 25–44; P. Pettit, *Republicanism: a theory of freedom and government* (Oxford, 1997); J. W. Maynor, *Republicanism in the Modern World* (Oxford, 2003), pp. 1–32; D. Weinstock, 'Introduction' to D. Weinstock and C. Nadeau, eds, *Republicanism: history, theory and practice* (London, 2004); D. Castiglione, 'Republicanism and its Legacy', *European Journal of Political Theory* 4 (2005), 453–65.
4. Robert Filmer, *Patriacha and Other Writings*, ed. P. Laslett (Oxford, 1949); John Locke, *Two Treatises of Government*, ed. and intro. P. Laslett (Cambridge, 1960).
5. H. Baron, *The Crisis of the Early Italian Renaissance: civic humanism and republican liberty in an age of classicism and tyranny* (Princeton, NJ, 1955, 2 Vols, revised 1966).
6. C. Robbins, *The Eighteenth-Century Commonwealthman: studies in the transition, development, and circumstances of English liberal thought from the Restoration of Charles II until the War with the Thirteen Colonies* (Cambridge, Mass., 1959).

7. F. Gilbert, *Machiavelli and Guicciardini: politics and history in sixteenth-century Florence* (Princeton, NJ, 1965), B. Bailyn, *The Ideological Origins of the American Revolution* (Cambridge, Mass., 1967); G. Wood, *The Creation of the American Republic, 1776–1987* (Chapel Hill, NC, 1969); J. Dunn, *The Political Thought of John Locke: an historical account of the argument of the 'Two Treatises of Government'* (Cambridge, 1969).

8. R. E. Shalhope, 'Towards a Republican Synthesis: the emergence of an understanding of republicanism in American history', *William and Mary Quarterly*, 3rd series, 29 (1972), 49–80; I. Kramnick, *Republicanism and Bourgeois Radicalism. Political ideology in late eighteenth-century England and America* (Ithaca, NY, 1990); M. P. Zuckert, *Natural Rights and the New Republicanism* (Princeton, NJ, 1994); J. Hankins, 'The "Baron Thesis" after Forty Years and Some Recent Studies of Leonardo Bruni', *Journal of the History of Ideas* 56 (1995), 324–30; W. J. Connell, 'The Republican Idea' in J. Hankins, ed., *Renaissance Civic Humanism* (Cambridge, 2000), pp. 14–29.

9. J. G. A. Pocock, 'The History of Political Thought: a methodological enquiry', *Philosophy, Politics, and Society*, 2nd series, ed. P. Laslett and W. G. Runciman (Oxford, 1962), pp. 183–202; J. Dunn, 'The Identity of the History of Ideas', *Philosophy* 43 (1968), 85–104; Q. R. D. Skinner, 'Meaning and Understanding in the History of Ideas', *History and Theory* 8 (1969), 3–53.

10. J. G. A. Pocock, 'Quentin Skinner. The history of politics and the politics of history', *Common Knowledge* 10 (2004), 532–50 (537–8).

11. On the differences between the approaches of Pocock and Skinner in the 1960s and subsequently, see I. Hampsher-Monk, 'Political Languages in Time: the work of J. G. A. Pocock', *British Journal of Political Science* 14 (1984), 159–74, and 'The History of Political Thought and the Political History of Thought' in Castiglione and Hampsher-Monk, *The History of Political Thought in National Context*, pp. 159–74.

12. S. Collini, 'General Introduction' in S. Collini, R. Whatmore and B. Young, eds, *Economy, Polity, Society: British intellectual history, 1750–1950* (Cambridge, 2000).

13. See especially J. Gunnell, *Political Theory: tradition and interpretation* (Cambridge, Mass., 1979), and 'Time and Interpretation: understanding concepts and conceptual change', *History of Political Thought* 19 (1998), 641–58; C. Condren, *The Status and Appraisal of Classic Texts* (Princeton, NJ, 1985); P. L. Janssen, 'Political Thought as Traditionary Action: the critical response to Skinner and Pocock', *History and Theory*, 24 (1985), 115–46; Tully, *Meaning and Context*; D. Harlan, 'Intellectual History and the Return of Literature', *American Historical Review* 94 (1989), 581–609; R. Tuck, 'History of Political Thought', in P. Burke, ed., *New Perspectives on Historical Writing* (Pennsylvania, 1992); M. Richter, 'Reconstructing the History of Political Languages: Pocock, Skinner and the Geschichtliche Grundbegriffe', *History and Theory* 29 (1990), 38–70, and *The History of Social and Political Contexts: a critical introduction* (New York, 1995); M. Bevir, *The Logic of the History of Ideas* (Cambridge, 2001); K. Palonen, *Quentin Skinner: history, politics, rhetoric* (Cambridge, 2003), pp. 29–60.

14. J. Coleman, 'The History of Political Thought in a Modern University', *History of Political Thought* 21 (2000), 152–71; D. Runciman, 'The History of political thought: the state of the discipline', *British Journal of Politics and International Relations* 3 (2001), 84–104; D. Castiglione and I. Hampsher-Monk, 'Introduction. The history of political thought and the national discourses of politics', S. Collini, 'Postscript. Disciplines, canons and publics: the history of "the history of political thought in comparative perspective"', in Castiglione and Hampsher-Monk, *The History of Political Thought in National Context*, pp. 1–9, 280–302; K. Haakonssen, 'The History of Eighteenth-Century Philosophy: history or philosophy?' in K. Haakonssen, ed., *The Cambridge History of Eighteenth-Century Philosophy* (Cambridge, 2006), pp. 3–25.

15. C. Robbins, review of *The Machiavellian Moment*, *William and Mary Quarterly*, 3rd series, 33 (1976), 335–7; J. Hexter, review of *The Machiavellian Moment*, *History and Theory* 16 (1977), 326–7; D. R. Kelley, review of *The Foundations of Modern Political Thought* 40 (1979), 666; W. J. Connell, 'The Republican Idea' in Hankins, *Renaissance Civic Humanism*, p. 26.

16. Q. Skinner, *Machiavelli* (Oxford, 1981); 'Machiavelli on the Maintenance of Liberty,' *Politics* 18 (1983), 3–15; 'The Idea of Negative Liberty: philosophical and historical perspectives', in R. Rorty, J. B. Schneewind and Q. Skinner, eds, *Philosophy in History* (Cambridge, 1984), pp. 193–221.

17. J. Pocock, *The Machiavellian Moment. Florentine political thought and the Atlantic republican tradition* (Princeton, NJ, 2003), pp. 83–330; Q. Skinner, *The Foundations of Modern Political Thought. Volume One. The Renaissance* (Cambridge, 1978), pp. 3–48, 69–112, 139–89.

18. Pocock, *The Machiavellian Moment*, 'Afterword', p. 554.

19. J. Pocock, 'Machiavelli and the Rethinking of History', *Il Pensiero Politico*, 27 (1994), 215–30.

20. Pocock, *The Machiavellian Moment*, pp. 401–22.

21. Ibid., 'Afterword', pp. 572–3.

22. Ibid., pp. 401–552; J. Pocock, 'Political Thought in the English-speaking Atlantic, 1760–1790', in J. G. A. Pocock, G. Schochet and L. G. Schwoerer, eds, *Varieties of British Political Thought, 1500–1800* (Cambridge, 1996), pp. 246–320.

23. Pocock, *The Machiavellian Moment*, p. 572.

24. J. Pocock, 'The Machiavellian Moment Revisited: a study in history and ideology', *Journal of Modern History* 53 (1981), 49–72; 'Between Gog and Magog: the republican thesis and the *Ideologia Americana*', *Journal of the History of Ideas* 48 (1987), 325–46.

25. H. F. Grœnvelt [Dumont, S. Romilly, J. Scarlett], *Letters containing an account of the late Revolution in France, and Observations on the Constitution, Laws, Manners, and Institutions of the English* (London, 1792), Letter 1, p. 11.

26. Ibid., Letter 11, p. 179.

27. Skinner published an article on Bolingbroke in 1974 ('The Principles and Practice of Opposition: the case of Bolingbroke versus Walpole' in N. McKendrick, ed., *Historical Perspectives*, London, pp. 93–128) but it was only with *Liberty before Liberalism* that he began to make general interpretative statements about intellectual life in the eighteenth century and beyond (see the revisions to 'The Case of Bolingbroke versus Walpole' published as 'Augustan Party Politics and Renaissance Constitutional Thought', in Q. Skinner *Visions of Politics. Volume 2: Renaissance Virtues* (Cambridge, 2002), pp. 344–67).

28. Skinner, *The Foundations*, pp. 54–6.

29. Pocock, *The Machiavellian Moment*, p. 558.

30. Ibid., pp. 559–60.

31. J. Pocock, *Barbarism and Religion, Vol. III: The First Decline and Fall* (Cambridge, 2003).

32. This point was first made by N. Tarcov, review of *The Machiavellian Moment*, *Political Science Quarterly* 91 (1976), 382.

33. Pocock, 'Afterword', *The Machiavellian Moment*, p. 568.

34. For examples, see M. Goldie, review of *Barbarism and Religion, Vols 1 and 2*, *Political Studies* 48 (2000), 1045; B. W. Young, review of *Barbarism and Religion Vol. 3*, *Albion* 36 (2004), 528–9.

35. Q. Skinner, 'John Milton and the Politics of Slavery' and 'Classical Liberty, Renaissance Translation, and the English Civil War' in Skinner, *Visions of Politics. Volume 2* pp. 286–307, 308–43.
36. Skinner, *Liberty before Liberalism*, pp. 26, 74.
37. Ibid., p. 33; Q. Skinner, *Machiavelli* (Oxford, 1981), pp. 56–73; Q. Skinner, 'Republican Virtues in an Age of Princes' and 'Machiavelli on Virtù and the Maintenance of Liberty' in Skinner, *Visions of Politics. Volume 2*, pp. 153–4, 173–4.
38. Skinner, *Liberty before Liberalism*, pp. 85–6, 17–18, 22–3. See also p. 55, n. 177.
39. Ibid., pp. 12, 21; Q. Skinner, *The Foundations of Modern Political Thought. Volume Two. The Age of Reformation* (Cambridge, 1978), pp. 302–48.
40. Q. Skinner, 'A Third Concept of Liberty', *Proceedings of the British Academy* 117 (2002), 237–68.
41. Skinner, *Liberty before Liberalism*, pp. 96–8.
42. Ibid., pp. 84–5.
43. Ibid., pp. 116–17.
44. See especially B. Worden, 'Factory of the Revolution', *London Review of Books* (5 February 1998), 13–15; I. Hampsher-Monk, review of *Liberty before Liberalism*, *Historical Journal*, 41 (1998), 1183–7; G. Newey, 'How do we Find Out what he Meant? Historical context and the autonomy of ideas in Quentin Skinner', *Times Literary Supplement* (26 June 1998), 29; P. Springborg, 'Republicanism, Freedom from Domination, and the Cambridge Contextual Historians', *Political Studies* 49 (2001), 851–77; P. A. Rahe, 'Situating Machiavelli' in Hankins, *Renaissance Humanism*, pp. 270–308; B. Fontana, 'In the Gardens of the Republic', *Times Literary Supplement* (11 July 2003); P. Zagorin, 'Republicanisms', *British Journal of the History of Philosophy* 11 (2003), 701–14; M. Albertone, 'Nuove discussioni sull'idea di repubblica nel XVIII secolo', *Rivista Storica Italiana* 114 (2002), 458–76; J. Moore, review note, *Political Studies Review* 2 (2003), 56–7; D. Wootton, review of *Republicanism: A Shared European Heritage*, *English Historical Review* 120 (2005), 135–9, and '"De Vera republica": the disciples of Baron and the Counter-Example of Venturi' in M. Albertone, ed., *Il Repubblicanesimo Moderno: L'idea di repubblica nella riflessione storica di Franco Venturi* (Naples, 2006); Pocock, 'Quentin Skinner', pp. 542–3.
45. J. C. D. Clark, *English Society, 1688–1832: ideology, social structure and political practice during the ancien regime* (Cambridge, 2000 [orig. 1985]), *Revolution and Rebellion: state and society in England in the seventeenth and eighteenth centuries* (Cambridge, 1986) and *The Language of Liberty 1660–1832: political discourse and social dynamics in the Anglo-American world* (Cambridge, 1994).
46. I. Hont, 'Introduction' and 'The Rhapsody of Public Debt: David Hume and voluntary state bankruptcy' in I. Hont, *The Jealousy of Trade* (Cambridge, Mass., 2005), pp. 1–156, 353–325; M. Sonenscher, 'The Nation's Debt and the Birth of the Modern Republic: the French fiscal deficit and the politics of the revolution of 1789', *History of Political Thought* 18 (1997), 64–103, 267–325.
47. J. Dunn, 'The Identity of the Bourgeois Liberal Republic' in B. Fontana, ed., *The Invention of the Modern Republic* (Cambridge, 1994), pp. 209–10.
48. J. Shklar, 'Montesquieu and the New Republicanism' in J. Shklar, *Political Thought and Political Thinkers* (Chicago, 1998), pp. 244–61.
49. D. Hume, 'Of Civil Liberty', *Political Essays*, ed. K. Haakonssen (Cambridge, 1994), p. 52.
50. Ibid., 55–7.
51. Hume, 'Of the Independency of Parliament' in *Political Essays*, p. 25: 'How much … would it have surprised such a genius as Cicero, or Tacitus, to have been told,

that, in a future age, there should arise a very regular system of *mixed* government, where the authority was so distributed, that one rank, whenever it pleased, might swallow up all the rest, and engross the whole power of the constitution. For so great is the natural ambition of men, that they are never satisfied with power; and if one order of men, by pursuing its own interest, can usurp upon every other order of men, it will certainly do so, and render itself, as far as possible, absolute and uncontrollable. But, in this opinion, experience shews that they would have been mistaken. For this is actually the case with the British constitution.'

52. Q. Skinner, 'States and the Freedom of Citizens' in Q. Skinner and B. Strath, eds, *States and Citizens. History, theory, prospects* (Cambridge, 2003), pp. 11–27.
53. R. Price, *Observations on the Importance of the American Revolution* (London, 1785), p. 72n, and *The Evidence for a Future Period of Improvement in the State of Mankind* (1787), *Political Writings*, ed. D. O. Thomas (Cambridge, 1991), p. 164.
54. G. E. Aylmer, review of *The Foundations of Modern Political Thought, English Historical Review* 95 (1980), 149.
55. F. Venturi, *Utopia and Reform in the Enlightenment* (Cambridge, 1971), pp. 18–19.
56. F. D'Ivernois et al., *An Historical View of the Constitution and Revolutions of Geneva in the Eighteenth Century* (London, 1784), p. 321.
57. Ibid., 149.
58. Ibid., 152.
59. E. Clavière and J.-P. Duroveray, *Pieces justificatives, pour messieurs Du Roveray & Clavière* (Geneva, 1780), pp. 52–3; F. D'Ivernois, 'La loi de la réélection, envisagée sous son vrai Point-de-vue', *Mémoires* (Geneva, 1780), p. 55.
60. M.-C. Pitassi, *De l'orthodoxie aux Lumières. Genève 1670–1737* (Geneva, 1992).
61. F. D'Ivernois, *Des Révolutions de France et de Genève* (London, 1795), *Coup-d'oeil sur les assignats: et sur l'état des finances et des ressources de la République Française, au 1er Janvier 1796* (London, 1796), and *Histoire de l'administration des finances de la république française, pendant l'année 1796* (London, 1797).
62. See especially D. Wootton, 'The Republican Tradition: from commonwealth to common sense', and B. Worden, 'Marchamont Nedham and the Beginnings of English Republicanism, 1649–1656' and 'Republicanism and the Restoration, 1660–1683' in D. Wootton, ed., *Republicanism, Liberty, and Commercial Society, 1649–1776* (Stanford, Calif., 1994), pp. 45–81, 139–98; B. Worden, *Roundhead Reputations. The English Civil Wars and the passions of posterity* (Harmondsworth, 2001), pp. 122–46, 209–10.

7

intellectual history and the history of science

james livesey

The relationship between intellectual history and the history of science is confused and confusing. Intellectual history is what history of science is trying hard not to be. Intellectual history claims that there is something specific about intellectual activity that calls for a programme of research that is in some important way different to the practice of political, social, economic or even cultural history. History of science, to the contrary, is organised around the claim that there is nothing specific to the sciences that cannot be understood through the common working practices of the historian. Science is a coherent and identifiable subject of study, but not essentially different to any other. These differing attitudes are perplexing since one might assume that the two fields would share a common interest in ideas. In fact, history of science and intellectual history diverge from one another precisely at this shared point of origin of attention to the claims humans have made to hold true accounts of themselves and the world. This divergence is complicated by the fact that it is unwitting. History of science and intellectual history find themselves at odds in consequence of positions held in other debates. If the two projects are to establish more cordial relations they need firstly to clarify their existing hostility.

i

To establish its credibility as a genuine field of research history of science has had to distinguish itself from accounts of their own history offered by scientists.[1] A powerful and intellectually respectable tradition holds that the history of science is science itself.[2] Science understands itself as an autonomous, self-correcting endeavour. The hypothetico-inductive method, it is argued, adequately explains the dynamic nature of science. Scientific realism holds that the results of scientific practice are objectively true, not in the sense that each statement of scientific fact is eternal, since one of the most obvious features of scientific practice is that every individual statement is open to revision. However, the development and even the revolutions in science do not require any sociological or cultural external impetus to the practice, the practice revolutionises itself. Science is genuinely progressive because later developments in science themselves explain how the earlier positions approximated to the revised version.[3] While one might give an historical or sociological account of the conditions that would allow this practice to emerge, in the manner of Robert Merton, the distinguishing feature of the practice remains its separation from other areas of human endeavour.[4] Thomas Kuhn's famous theory of scientific revolutions, which is often misunderstood as opening a door to a socially embedded history of science, in fact does exactly the opposite.[5] His argument holds that background assumptions that construct particular views of reality are eventually overturned by the accumulation of incommensurabilities, phenomena which could not be reconciled to the currently dominant theories. Scientists' anxieties about some anti-realist premises in Kuhn's argument about incommensurability, such as his contention that scientists in separated cultural traditions could live in worlds with differently constructed natural objects generated by their differing background cultural assumptions, blinded them to his commitment to a fundamentally orthodox vision of anything worthy of being termed a science as autonomous and driven by the process of testing hypotheses. The orthodox history of science is so powerful that it even captures the heretics. Culture might drive theory-selection in Kuhn's account, but nature, in the long run, rendered theories successful or unsuccessful.

If the history of science is science itself then the discipline of the history of science creates no new knowledge. The discipline of the history of science, understood in this way, merely glosses the process of scientific progress or illustrates it with biography. The bodies of work on Newton's alchemy, or his theological speculations, for instance, make for interesting insights into Newton's personality, but have no more purchase on his place in the history of science than his political opinions or his sartorial taste. All of these activities were not falsifiable and to consider them as elements of a science would be to make a categorical error – and error, in science, has no rights. History of science might have a domain in the exploration of scientism, the circulation of ideas derived from science in politics or society, such as the history of Social Darwinism. History of science would only explain something, add something to the history of Darwinism implicit in the practice of the life sciences, to the extent that Social Darwinism was a misapplication of Darwin's ideas.[6] The task of history of science then becomes to police the boundaries of science and to point out the perils of misapplication, such as scientific racism.[7] Again the autonomous discipline of the history of science is granted the domain of error for its own, but the practice of science remains inviolate.

History of science established its autonomy from the authority of science by transgressing the boundary between truth and error. Scholars such as Richard Westfall and Betty Jo Tweeter Dobbs argue that to understand Newton's science one has to approach it through his theology and his alchemy, that there is no division in his intellectual life between science and other activities.[8] The declaration of independence for history of science is the symmetry principle. The symmetry principle argues that in the history of science the same categories and elements must be used to interpret and explain true and false scientific programmes, that there are no different methods appropriate to study phrenology, astrology and psychology, or even alchemy and chemistry. The first corollary of the symmetry principle is that the truth or falsity of any statement or theory is irrelevant to its history. This is a neat, formal distinction between two orders of explanation that was already inherent in such classic formulations as Popper's division between the context of discovery and the context of validation. The clever inversion that created a genuine field of history of science was to prioritise the former over the latter. The symmetry principle is not a position in epistemology; rather it is a methodological postulate that asserts neutrality about the validity of any particular scientific position, programme of research or even entire discipline. It really asserts a division of labour, arguing that while only scientists are qualified to assert the validity of a proposition, only historians are competent to give a good account of its discovery, circulation and importance. The territory of the historian of science is to explain not why something is true but why it mattered, why it was meaningful to a particular time and place.

Such perfect neutrality on epistemology cannot be sustained by history of science when it has to establish its own rules and procedures. The symmetry principle only generates a really autonomous field of history of science if it can derive a rule of method. This second derivation from the symmetry principle is the more threatening to scientific orthodoxy. It argues that science is a social activity, like any other. In consequence, the same categories used to illuminate any other social practice, such as class, gender, race or nationality, should be used in the historical investigation of scientific practice. The 'strong programme', associated with the Edinburgh University Science Studies Unit, pursued a wholehearted sociology of science and argued that the truth value of particular scientific ideas were themselves social in origin, thus collapsing the discovery/validation dichotomy.[9] The scandalised reaction of practising scientists to the attack on their claims to represent reality led to the outbreak of the 'science wars' which are with us still.[10]

While the philosophical consequences of social constructivism, the claim that truth is made not found, are fascinating, the more practical consequence for doing history was that wide new areas of analysis were opened up for a liberated history of science. Andrew Pickering's *Constructing Quarks* is a controversial account of the development of high energy physics in the post-war era, and a good example of the value of the 'strong programme'.[11] Pickering avoids the philosophical problem of scientific realism by addressing not what the scientists in the high energy physics community thought about what they were doing, their theoretical reconstructions, but what they actually did, their laboratory behaviour. He analyses the choices they made to favour one set of results over another, to pursue a particular line of enquiry or endorse particular theoretical avenues, in terms of resources available, the congruence of interests and the compatibility or otherwise of proposed activities with existing traditions and institutions, all strategies of analysis familiar to historians in many other contexts. Without denying the specific identity of the social group he addressed himself to, and its control of very esoteric skills, he established that they could be understood through classic historical technique. The eye-catching claim that quarks were invented should not really have been that surprising since the original quark is found in *Finnegans Wake* and in physics is a theoretical construct postulated in response to experimental results, made by machines not found or directly observed in nature. The really important consequence of Pickering's work was that no area of science, no matter how arcane, was beyond the purview of the historian of science, who would analyse it as just another social practice. This social realist angle of analysis has particular purchase on sciences, such as economics, that derive their autonomy by analogy with the natural sciences. Philip Mirowski's work reconstructing the accumulation of interests that created modern economics has been particularly interesting in this regard.[12]

Treating communities of scientists like any other craft or trade group, with their own entry conditions, dominant self-understandings, rituals, work practices and politics, has generated a powerful set of tools for history of science.[13] The benefits have been visible at every level of analysis. Redirecting attention away from the history of cognitive values, a domain within the history of philosophy, to the history of networks, has widened the scope of history of science to include subaltern practitioners. This has transformed our understanding of the eighteenth- and nineteenth-century life sciences in particular. Released from the straitjacket of precursors to Darwin, we now understand this world to have been full of roads untaken, such as Reamur's anthropomorphic typology, and of different characters, such as female, artisan and peasant botanists.[14] The theme of this subject has ceased to be how it attained professional respectability and theoretical integration, but instead emphasises its variety, complexity and fragility. This stance also opens up the question of the relationship of developments in scientific communities to other social processes, such as class formation and industrialisation. Scholars such as Thackery, Inkster and Stewart have illustrated how provincial scientific communities in England contributed to the creation of a distinctive provincial culture and economy.[15] The symmetry principle, by bracketing the feature of their work most important to scientists, its truth, paradoxically allows historians of science to generate a more full, detailed and compelling account of the place of the various sciences in wider historical context. The explanatory strategy of historians of science has been to argue that their object of interest is a quotidian practice which needs no special technique or approach to be understood. Intellectual historians, on the other hand, sharply distinguish between their domain and that of the quotidian. The central commitment of intellectual history is to a non-deterministic and non-reductionist account of thought and its expression. Thought is conceptualised in intellectual history as representation and expression conducted by a subject. The subject, author of thought, is the key to this programme.[16] All intellectual activity has a context, and it is through that context that the historian can reconstruct the meaning of the acts of expression, but those contexts do not determine, merely explicate. The qualities associated with thought, such as insight, creativity and novelty, are all characteristics of subjects, authors of their own ideas. Subjects are free, in the sense of being the irreducible agents of the acts of thought, and therefore intellectual history is the history of human freedom.[17] Through communicating thought human societies can create their contexts and themselves and so escape from the determinations of nature. In its most ambitious formulations intellectual history offers a programme for the whole discipline: history as a whole can be understood as the action of human willing subjects and intellectual history prioritised since it captures that free action.

In consequence the practice of intellectual history cannot be to explicate the strategy of a thinker or group in functionalist terms. What counts as thought is not strategic, not because it does not address its political, social or economic context, but because it is not determined by such context.[18] Intellectual history is a thoroughly hermeneutic exercise, participation in the conversation of mankind, in Oakeshott's phrase, and that image of intellectual history as a discipline that makes its practitioners fit conversationalists is one that recurs.[19] Quentin Skinner's methodological principle that the account of a thinker has to be one that would be acceptable to the person being analysed is effectively another way of claiming that the intellectual historian is characterised by enormous hermeneutic sensitivity and by respect for the thinker at hand.[20] A properly trained intellectual historian would be one who could leave the limitations and determinations of the quotidian and respond creatively and appropriately to the ideas of his or her subject.

Obviously any and every version of the sociology of knowledge is antithetical to this vision of intellectual history and every version of the history of science, no matter what the divergences among them, is committed to sociology of knowledge. This is not the only axis of opposition between intellectual history and the history of science. Intellectual history and the orthodox account of science and its history are collusive with one another, behind their backs as it were. The Kantian division of intellectual labour between understanding and explanation perfectly describes the relationship between 'the two cultures'.

While the lost unity of knowledge might be regretted, the demarcation between the sciences and the humanities has been extremely useful since it has meant that both realms had a clear border that they could respect. It is noticeable that the history of Freudianism, which was the exception that did not respect this boundary, has attracted scholars who were suspicious of this neat distinction and the exploration of the psychoanalytic tradition continues to be a disturbing and idiosyncratic area of intellectual history. History of science disturbs the treaty of demarcation between these two realms of knowledge and so threatens the integrity of both. The very existence of history of science advertises the limits of ambition of intellectual history in this regard.

Another barrier between the history of science and intellectual history has been erected through the inheritance of a prejudice from German cultural criticism by intellectual history. Intellectual history's embrace of the ideal of the historical subject creating their world through thought and expression does not commit the field to any particular kind of thought and expression. As Slavoj Žižek has asserted the subject is a disturbing idea that is not consistently available for use as the basis for any claim to authority.[21] Just what constitutes the subject is a question that is far from settled. However, in practice the genealogy of intellectual history has committed it to a very particular vision of authentic subjectivity. The inspirations for Anglophone intellectual history in Europe were Collingwood and Oakeshott and they inherited their notion of what counted as thought from their reading of conservative German scholars of the 1930s, Otto Bruner, Carl Schmitt and Hans Freyer. Carl Schmitt's work also inspired the tradition of intellectual history associated with Leo Strauss and Hannah Arendt's notion of politics as tragedy inspired the work of J. G. A. Pocock. All of these research programmes share roots in Heidegger's critique of modernity, a critique that was particularly focused on science and technology. For Heidegger, authentic thought was the antithesis of science. Authentic thought was above all else a disposition, openness to Being that was the opposite to the instrumental reason he argued characterised the scientific method. Heidegger's indictment of scientific rationality was not restricted to post-Copernican Western science; rather he saw it implicit in the history of metaphysics since Aristotle.[22] However, he discerned the fragments of an exit from the onto-theological in the classical tradition and this presumption in favour of classical scholarship has been inherited by his historian progeny. In consequence intellectual history has been dominated by the history of humanism and the humanist intellectual of the seventeenth century remains the enduring image of authentic intellectual activity. Nor is this prejudice restricted to scholars of a conservative cast of mind; the *Dialectic of Enlightenment* introduced the same themes to the Marxist tradition. The consequence for intellectual history has been to privilege the study of humanist culture, of classical scholarship, and to be strangely blind to the creativity of the thought of the scientific tradition. The authentic thinker, for intellectual history, remains the humanistic scholar.

History of science and intellectual history do not have an engaged relationship, rather they both suffer under a misunderstanding. From the perspective of the history of science intellectual history can appear idealist. Great thinkers, who seem to be restricted to humanist scholars of the early modern period, are free subjects, but their thought seems to condition the existence of everyone else. The use of Rousseau to understand the history of the French Revolution or the privilege accorded to Hobbes's account of sovereignty in the evolution of the state are examples of currents in intellectual history opposite to trends in the history of science which seek to reveal the social processes that produce the authored theory: the artisans making the instruments, the patrons generating credibility, the church authorities endorsing the theological implications of the idea. Intellectual history can also appear methodologically antique. For all the ink that has been spilled on speech act theory and the illocutionary and perlocutionary force of particular utterances, the methodology of intellectual history has remained thoroughly philological and the critique of historical philology has been the basis for every variety of social history.[23] The first example of Cambridge School intellectual history was Peter Laslett's edition of Locke's *Two Treatises on Government*.[24] The import of this text was to argue that intellectual history had been impoverished by taking ideas for objects distinct from texts. By concentrating on texts and their contexts historians could reconstruct the import of the act of writing and publishing, restoring history to the history of ideas. Laslett's discovery that Locke's work was written as in exile in the Netherlands as a justification for tyrannicide, and not in the aftermath of the Glorious Revolution as a charter for a property-owning elite, was a brilliant justification for his claims. Close textual analysis of this sort is not entirely foreign to history of science; a good example would be Rose-Mary Sargent's book on Robert Boyle.[25] However, the ambition of intellectual history to participate in and mimic the methods of its humanist subjects is not generally shared.

From the other shore, history of science would seem a curiously self-contradictory enterprise. The effort invested in denaturing science, at obscuring its specificity and denying its power to reveal the nature of the universe seems disproportionate to the modesty of the methodological claims advanced. History of science's relationship to the many varieties of science studies is not entirely clear, and while history of science might well claim epistemological neutrality, practitioners of science studies, from Rorty to Foucault, do not. A particular version of history of science has been accused of being ideological, of really being a Trojan horse for postmodernism.[26] While postmodernism has attracted attention from historians in some fields such as the history of sexuality or of post-colonialism, in general its presuppositions have been seen to be so antithetical to any coherent practice of history that it has been more ignored than contested.[27] For the majority of intellectual historians these concerns are very uninteresting. Finally, the symmetry principle seems to commit history of science to the historical sin of analepsis; by bracketing the truth of any scientific idea historians of science generate an account of scientific practice that could not be endorsed by its practitioners. This kind of historian seems to ignore a key condition of the practice as understood by the practitioner. This kind of ideology critique, of claiming a special knowledge that reveals what historical actors were *really* doing despite what they might say to the contrary, is anathema to intellectual history.

This unsatisfactory interaction between two neighbouring fields of enquiry lacks spirit and passion. Each displays a mild antipathy to the other, but neither is engaged enough with the other's concerns for antipathy to escalate into anything as creative as sustained criticism. The absence of a healthy engagement between these two fields has been very damaging to both and has impoverished the contiguous fields of social and political history by making it very difficult to generate integrated histories in which the manner in which intellectual and scientific cultural resources were deployed can be narrated convincingly. It is increasingly clear that the history of knowledge, information and communication is an important topic through which we can hope to renew our understanding of the past and in particular of the early modern period.[28] Paul Slack even claims that the subject has become so fashionable that the more mundane aspects of communication, such as interaction between government offices, may be lost to view.[29] Yet despite this plethora of interesting work the absence of a critical engagement between the two sub-disciplines most engaged with the history of justified belief, of authoritative knowledge, or even of heterodox opinion, has hampered the efforts of various historians to move forward on this front. The related history of the emergence of laboratory science and the political theory of the English Civil War is one such area where there has been a lack of integration as has the relationship between the intellectual history of the French Revolution and the emergence of state-organised and validated social and life sciences. It is through examining these problems that we get a real sense of what has been lost through this lack of engagement, and some hints of where such an engagement might be encouraged.

ii

Around 1660, writes Ian Hacking, laboratory science began, and he goes on to claim that this is Western Europe's chief contribution to world history.[30] Hacking advances this contention in his review of the second major monograph to emerge from the research programme of Steven Shapin and Simon Shaffer.[31] Shapin and Shaffer have offered an exciting answer to one of the most important questions in history – how did a new kind of authoritative knowledge, natural science, emerge from the old natural philosophy in the seventeenth century? The shape of what had happened has long been clear.[32] Knowledge had been knowledge of Aristotle's four causes: material, efficient, formal and final. Knowledge was knowledge of universals; particular instances could not be the object of genuine knowledge since they presented themselves through accidents, such as colour, rather than essential qualities, such as mass. Real knowledge could only be general since particular knowledge was just that, local and specific and so conditional. Truth was, by definition, unconditional.

The intellectual revolution of the seventeenth century was the overturn of this hierarchy of the universal in favour of the specific, be it the clear and distinct idea beloved of Descartes, or Bacon's facts. One of the most obvious consequences of this transvaluation was the abandonment of teleology as a strategy for having knowledge of motion, in favour of mechanism.[33] The cosmological frame that guaranteed the certainty of causal knowledge was shattered. Now knowledge had to be built up out of particular elements, and figures such as Descartes and Bacon consciously emphasised that the idea of knowledge would have to be reconstructed on this new basis since the Aristotelian foundations had proven so insecure. The image of the mosaic replaced that of harmony as a representation of the world of knowledge, and early idealists of the new practice, especially Bacon, pointed the way toward a reconstruction of that mosaic as a collective and collaborative project of experiment.[34] There were other salient aspects to the changes in the world of knowledge, such as the new importance of mathematics as a means of manipulating and organising particulars, but the most important change was the commitment to particulars.[35]

Shapin and Shaffer recapture just how strange and alarming it was to reconceptualise what counted as knowledge in their analysis of Thomas Hobbes's critique of the experimental programme. Hobbes wrote his *Physical Dialogue of the Nature of the Air: a conjecture taken up from experiments recently made in London at Gresham College*, originally in Latin, in response to Robert Boyle's *New Experiments Physico-Mechanical* of 1660. Hobbes was no Aristotelian conservative, but was alive to the possible intellectual and political consequences of Boyle's work, and in particular to his experiments on gases that hypothesised a vacuum in order to derive a value for air pressure. What Boyle entertained as an experimental hypothesis Hobbes rejected as a theological danger. A vacuum implied a space which was not filled with matter, and in consequence one that might be filled with 'ether' or some other spiritual substance. For Hobbes the idea that there might be two principles of reality (matter and spirit) was the ontological basis from which separate claims to temporal and spiritual authority might be made and the consequence would be dissension and civil conflict. Hobbes did not restrict himself to a jeremiad about the possible consequences of a particular experimental result, he went on to offer a sustained critique of the experimental programme for knowledge. He identified the problem of experimental regress: the authority claims of the experimenters rest on accounts of results from procedures that the experimenters themselves had created. 'Torturing nature' in the air pump set up two problems of interpretation: that of assessing if it was properly working and then of assessing what phenomena it exhibited when it was working. Reproducibility did not solve this problem of arcane facts, since the criterion of reproducibility was reproduction of the effects that Boyle had produced. Hobbes went on to argue that the warrant offered for the generality of scientific fact, the public interrogation of method and claim, was fictive. The Royal Society, where Boyle had conducted his experiments, was not public but a closed corporation; in fact, a perfect example of the subsidiary body that threatened the peace of the state. In response to his critique of laboratory science, Hobbes offered an alternative model of a science based on deductive inference from axioms, based on the model of geometry. What Shapin and Shaffer identify is a dispute about the form of authoritative knowledge in what was quite literally a revolutionary moment, one in which not only the institutions of knowledge were at stake.

This revolutionary moment is also the subject of Quentin Skinner's analysis of Hobbes's civil science or political theory.[36] Skinner entirely concurs with the account of Hobbes's model of a science offered by Shapin and Shaffer: for Hobbes, a science is a deductive array of axioms properly derived. It is certain because the initial definitions are beyond doubt. In the *Leviathan* he explains that such definitions, of justice, liberty, and so on, are apodictic because enunciated by the sovereign. It is rhetorically compelling because it is promulgated in the plain, unadorned oratory of truth. Given the importance of Hobbes to the construction of such central concepts of political life as sovereignty and rights one might expect a fruitful analysis of his epistemology, of the consequences of the position he took up in the debate with the experimentalists of the Royal Society. This is what Skinner explicitly rejects. In the opening of the second part of the book he acknowledges Hobbes's participation in scientific controversy, but comes to the conclusion that this is simply not relevant to understanding his civil science: 'we find that his intellectual formation was overwhelmingly indebted not to the culture of science but rather to the humanist literary culture'.[37] Skinner argues that Hobbes's revolution in civil science was conducted in dialogue with the rhetorical tradition of the humanists rather than with the new natural philosophy. Skinner deploys exactly the distinction between understanding and explanation that polices the division between science and the humanities in the contemporary academy, to distance Hobbes's concerns from those of figures such as Boyle. Skinner hypothesises two distinct traditions of enquiry, rhetorical and scientific: one committed to the explication and guidance of beings that are fundamentally teleological, and the other committed to mechanistic accounts of the relations of natural objects.[38] Hobbes's concern, he argues, is with the first, and the particular meaning he gives to 'civil science' reflects that. A debate that might occur, about Hobbes's embrace of a particular idea of scientific authority in a moment of reconstruction of authoritative knowledge, does not take place and the history is impoverished as a result. Hobbes's notion of a science is so far from the model of experimental science that animated philosophical radicals from Spinoza through Bayle to Diderot, and so obviously relates to the very different political commitments of the traditions, that the interaction between the two models cries out for analysis.[39] By insulating Hobbes's ideas about civil science within the humanistic tradition, Skinner distances them from the discussions of and experiments on nature that drove much political speculation and experimentation.

It is regrettable that a possible discussion of the interactions of contrasting ideals of authority and knowledge does not occur, but it is also understandable. Skinner wants to give an account of Hobbes's search for a robust, true civil science that respected Hobbes's contribution of 'teaching philosophy to speak English'. Shapin especially argues that the truth norms of early modern experimental science were derivative of social norms of gentlemanly credibility. In his *Social History of Truth*, he argues that the problem of verifiable but arcane facts, generated in voyages of exploration, laboratories and microscopic observations, was general and radical.[40] Just how was a reader or listener to distinguish a good account of an observed account from a bad, especially when these were accounts of unusual experiences? Reconstructing the mosaic of nature was impossible without solid tesserae. Shapin argues this problem was solved by moving attention from the testimony to the testifier: gentlemen were, by definition, able to make these judgements, thus one deferred to the judgement of gentlemen. Even as technically proficient an experimenter as Robert Hooke had less epistemic authority than a gentleman because of his social position, as a servant, if a highly valued one, of Boyle and the Royal Society.[41] Where Skinner sees the lodestar of apodictic certainty as the norm which allows Hobbes to create a new insight into the demands of political community, Shapin sees truth as a weak ideal that has to be supplemented by social authority if it is to organise and explain the multiplicity of observations of the world and nature. As Ian Hacking has perceptively pointed out, Shapin gives an account of trust as an alternative to an idea of truth. Gentlemanly decorum creates a space in which conventions could govern dispute about testimony, effectively collapsing the distinction between the scholar and the gentleman. Skinner's Hobbes, on the other hand, retains his distinctive contribution to the making of history, even if he does so at the price of being restrictively defined into a particular philosophical tradition. Hobbes's status as a servant of the Cavendish family is simply beside the point; he is an agent, author of his own thought, because he seeks the truth.

Social history of natural science and intellectual history of political science lose an opportunity to interrogate one another in this Hobbesian moment. Driven by antithetical models of the humanist and the scientist, the history of truth, knowledge and representations gets lost.[42] The loss is greater since this reinforces the estrangement between the fields on a topic where other work is bringing them together. For instance, Shapin was aware that his model of gentlemanly decorum as the guarantor of truth created a special problem of accounting for the witness and expertise of technicians – the men, like Denis Papin, who actually worked the instruments which created experimental knowledge. He solved this problem for himself by arguing that they were included in the model of the household. Experiment was understood as a domestic enterprise.[43] This clever resolution of a problem of membership in the scientific world has been unravelled by Rob Iliffe.[44] Iliffe's analysis of Robert Hooke's diaries reveals that the coffee house rather than the closed environment of the Royal Society was the space of interrogation of natural science and of the instruments through which it was demonstrated. Most importantly, the coffee house was the locus where gentlemanly priority was least salient and in which social order was established through rational discourse rather than the other way around.[45] Negotiation in the coffee house, between cultures and across social distinctions, demanded a common culture of truth, one that was robust enough to survive social difference. Hooke's collaborators were not the only persons who were credible witnesses but not under gentlemanly authority. Samuel Sorbière, for one, played a vital role as the intermediary between French and English natural science.[46] Sorbière was more broker than servant. Hobbes's desire for a true rhetoric was not essentially different from the natural philosophers' need for a way of speaking the truth in public.[47] The re-emergence of the horizon of truth as an independent and irreducible element in the history of knowledge is not the only point at which the distinction between the humanist and the natural philosopher is becoming difficult to sustain. Anthony Grafton's work has underlined that the humanist text was yet another species of the particular, along with the idea, beloved of Descartes, and Bacon's fact.[48] Analysis and rhetoric cannot be distributed as cognitive strategies between science and humanism. As Ann Blair has pointed out, the basic problem of too many particulars and the need for new categorical schemes to organise them plagued every kind of scholar.[49] Texts were the tesserae for another kind of mosaic. The working practices of the humanist scholar and of the natural philosopher were not distinct enough to sustain separate identities.

History of science and intellectual history could and probably should contribute to a history of truth in the English seventeenth century. The idea of truth was obviously central to a variety of actors, from religious radicals claiming divine illumination, through Issac Newton trying to discover the age of the earth as well as the mathematics of the celestial system, to Geoffrey King seeking to establish the mortality rate in London or William Petty seeking to understand the variety of resources that might be developed in Ireland. The truth was prized because it was a solid foundation on which a conflict-ridden and unstable polity might be reconstructed. However, the rift between history of science and intellectual history makes it very difficult to conceptualise that history, despite the fact that scholarship in contiguous areas renders the distinction between the two fields very difficult to justify. This is not a problem that is particular to the history of seventeenth-century England; a similar issue troubles the history of the French Revolution. The relevance of the projects of both intellectual history and history of science to the Revolution are obvious: it is the locus and object of political theory for the last 200 years and it was the period that created national scientific institutions and the social identity of the scientist. There is no lack of empirical examples of the interaction of these two themes either, from the anti-Newtonism of the popular radical Marat to the speculations on probability and voting by Condorcet. This site of the relationship of knowledge, intellect and power has inspired some outstanding monographs, such as Keith Baker's work on Condorcet, Charles Gillespie's on the political organisation of science and Ken Alder's on military engineers.[50] Yet despite these and other works no uniting paradigm through which the interaction of practices of knowledge and revolution might be addressed has emerged. Scholars have flirted with the question of technocracy and bureaucracy, especially in regard to political economy. However, no one has opened up a way of approaching the obvious importance of new institutions of knowledge, organised in institutes and practised in homes, that characterised the Revolution.[51] In the absence of such a research programme the subject continues to be illuminated through the study of political theory while the history of science addresses determining institutions of control. Again, the contested but vital value of truth is lost between the two research agendas and efforts at comparison end up instrumentalising one sphere in terms of the other. Sean Quinlan's comprehensive effort at an overview of the place of scientific authority in the Revolution is a good example here.[52] Quinlan reads scientific programmes entirely in terms of political projects of control. In his conclusion he ends up asserting that every scientist engaged in the project of reconstructing structures of enquiry and investigation became an instrument of 'authoritarian political solutions'.

iii

The interaction of intellectual history, particularly the history of political thought, with history of science, has been so fraught with misunderstanding and incomprehension that one is tempted to abandon the terrain altogether. In the face of theoretical reconstructions that obscure more than they illuminate, Margaret Jacob has strongly argued for a highly empirical model of history of science, using the history of national institutions and traditions of enquiry as the point of comparison.[53] A more ambitious and creative suggestion has emerged from feminist studies of science. Karen Barad has argued that history of science, and science studies more generally, is in thrall to a *question mal posée*. Historians should not have to choose between social constructivism, which denies all agency, and scientific realism, which asserts an impossibly pristine model of agency. Instead, she argues for 'agential realism' which 'theorises agency in a way that acknowledges that there is a sense in which the world "kicks back" (i.e. non-human and cyborgian forms of agency in addition to human ones) without assuming some innocent, symmetrical form of interaction between knower and known'.[54] Barad suggests a way in which the concern of the historian of science for context and that of the intellectual historian for agency can be reconciled, by broadening the conception of agency. This theoretical move is reinforced by empirical work criticising the very notion of an absolute form of agency, of the scientist who gave meaning to a brute reality. The feminist account of how the disenchantment of the world went in parallel with the feminisation of nature has underscored the extent to which the idea of agency that animates intellectual history is a disguised form of masculinity.[55] Lorraine Daston in turn has drawn the conclusion that history of science need not fear the subject, or agent, if that agent is not characterised by 'a hyperrational logic of trust and belief'.[56] History of science can be written as the search for truth by agents, but only if we revise our idea of agency and so better understand the truth they were concerned with. What such a renewed history of science would look like has been pioneered in the work of Bruno Latour and Donna Haraway. Latour's claim that 'we have never been modern' and Haraway's analysis of the interdependency of humans and non-humans as cyborgs both assert the same point: that the distinction between nature and culture has always been ideological and cannot be used as a strong analytic position.[57] These positions radicalise the study of practice by widening the range of elements that act, that have agency, and by seeing the production of truth as an emergent negotiation between human and non-human, rather than as the addition of specifically human quality to an otherwise cognitively inert world. These positions are controversial, since they are counter-intuitive and at such odds with the everyday understandings of scientists and indeed humanists. However, the capacious model of agency they offer seems to be the one approach that can reconcile history of science to intellectual history.

notes

1. See, for instance, Steve Woolgar, *Science: the very idea* (London, 1988).
2. Karl R. Popper, *Conjectures and Refutations: the growth of scientific knowledge* (2nd edn, London, 1965).
3. Paul M. Churchland, *A Neurocomptational Perspective: the nature of mind and the structure of science* (Cambridge, 1985); Clifford A. Hooker, *Reason, Regulation and Realms: toward a regulatory systems theory of reason and evolutionary epistemology* (New York, 1995); Philip Kitcher, *The Advancement of Science: science without legend, objectivity without illusions* (New York, 1993); Hilary Putnam, *The Many Faces of Realism* (La Salle, 1988).
4. Robert Merton, *On Social Structure and Science* (Chicago, Ill., 1996). The classic Mertonian case study is Robert Merton, *Science, Technology and Society in Seventeenth-Century England* (New Jersey, 1978 orig. 1938).
5. Thomas S. Kuhn, *The Structure of Scientific Revolutions* (3rd edn, Chicago, Ill., 1996); Steve Fuller, *Kuhn versus Popper: a struggle for the soul of science* (Cambridge, 2003).

6. In this style, see Peter J. Bowler, *The Non-Darwinian Revolution: reinterpreting an historical myth* (Baltimore, Md, 1988) or Richard Hofstadter, *Social Darwinism in American Thought* (2nd edn, New York, 1959).

7. Ronald E. Hall, *The Scientific Fallacy and the Political Misuse of the Concept of Race* (Lewiston, NY, 2004).

8. Betty Jo Tweeter Dobbs, *The Foundations of Newton's Alchemy or 'The Hunting of the Greene Lyon'* (Cambridge, 1975); Betty Jo Tweeter Dobbs, *The Janus Face of Genius: the role of alchemy in Newton's thought* (Cambridge, 1991); Richard Westfall, *The Life of Isaac Newton* (Cambridge, 1993); J. Force, 'Newton's God of Dominion: the unity of Newton's theological, scientific and political thought', in J. Force and R. H. Popkin, eds, *Newton and Religion: context, nature and influence*, International Archives of the History of Ideas 129 (Dordrecht and Boston, Mass., 1999), pp. 76–102.

9. For accounts of this programme, see Barry Barnes, *Scientific Knowledge and Sociological Theory* (Boston, Md, 1974); David Bloor, *Knowledge and Social Imagery* (Chicago, Ill., 1976); Barry Barnes, David Bloor and J. Henry, *Scientific Knowledge: a sociological analysis* (Chicago, Ill., 1996).

10. For a measured introduction to the 'science wars', see Jay A. Labinger and Harry Collins, eds, *The One Culture? A conversation about science* (Chicago, Ill., 2001).

11. Andrew Pickering, *Constructing Quarks: a sociological history of particle physics* (Edinburgh, 1974). For a discussion of Pickering's importance, see Ian Hacking, *The Social Construction of What?* (Cambridge, Mass., 1999), pp. 68–73.

12. Philip Mirowski, 'The Rhetoric of Modern Economics', *History of the Human Sciences* 3: 2 (1990), 243–57; Philip Mirowski, *Machine Dreams: economics becomes a cyborg science* (Cambridge, 2002).

13. Bruno Latour and Steve Woolgar, *Science in Action: how to follow scientists and engineers through society* (Philadelphia, Penn., 1987).

14. E. C. Spary, 'The "Nature" of Enlightenment', in William Clark, Jan Golinski and Simon Schaffer, eds, *The Sciences in Enlightened Europe* (Chicago, Ill., 1999), pp. 272–304; Anne Secord, 'Science in the Pub: artisan botanists in early nineteenth-century Lancashire', *History of Science* 32 (1994), 269–315; Ann B. Shteir, *Cultivating Women, Cultivating Science: Flora's daughters and botany in England, 1760–1860* (Baltimore, Md, 1996); James Livesey, 'Botany and Provincial Enlightenment in Montpellier: Antoine Banal père et fils 1750–1800', *History of Science* 43 (2005), 57–76.

15. Ian Inkster, *Scientific Culture and Urbanisation in Industrialising Britain* (Aldershot, 1997); A. Thackery, 'Natural Knowledge in Cultural Context: the Manchester model', *American Historical Review* 79 (1974), 672–709; Larry R. Stewart, *The Rise of Public Science: rhetoric, technology and natural philosophy in Newtonian Britain 1660–1750* (Cambridge, 1992).

16. For a criticism of this notion of thought as a species of purposive action, see Mark Bevir, 'Mind and Method in the History of Ideas', *History and Theory* 36: 2 (May 1997).

17. Benedetto Croce, *History as the Story of Liberty* (London, 1941), trans. Sylvia Sprigge, develops perhaps the clearest defence of this version of history.

18. James Tully, 'The Pen is a Mighty Sword: Quentin Skinner's analysis of politics', *British Journal of Political Science* 13: 4 (1983), 489–509.

19. Michael Oakeshott, 'The Voice of Poetry in the Conversation of Mankind' in *Rationalism in Politics and other Essays* (London, 1962), pp. 197–247.

20. Quentin Skinner, 'Meaning and Understanding in the History of Ideas', *History and Theory* 8 (1968), 3–53.

21. Slavoj Žižek, *The Ticklish Subject: the absent centre of political ontology* (London, 1999).
22. Martin Heidegger, 'The Question Concerning Technology', in David Krell ed., *Basic Writings* (London, 1978), pp. 307–42. Heidegger of course condemns 'subject-centred reason', and the subject of intellectual history derived from him is understood as *Dasein*.
23. François Simiand, 'Méthode historique et science sociale', *Revue de Synthése* 6 (1903), 1–22, 129–57.
24. John Locke, *Two Treatises of Government*, ed. Peter Laslett (Cambridge, 1988 [orig. 1963]).
25. Rose-Mary Sargent, *The Diffident Naturalist: Robert Boyle and the philosophy of experiment* (Chicago, Ill., 1995).
26. Margaret C. Jacob, 'Reflections on the Ideological Meanings of Western Science from Boyle and Newton to the Postmodernists', *History of Science* 33: 3 (1995), 333–57.
27. For examples of an engagement with postmodernism, see Derek Attridge, Geoff Bennington and Robert Young, eds, *Post-Structuralism and the Question of History* (Cambridge, 1987); Hayden White, *Metahistory: the historical imagination in nineteenth-century Europe* (Baltimore, Md, 1975); Patrick Joyce, *Democratic Subjects* (Cambridge, 1994). For post-colonial approaches, see Homi K. Bhaba, ed., *Nation and Narration* (London, 1990); Gayatri Spivak, *A Critique of Post-Colonial Reason: toward a history of the vanishing present* (Cambridge, Mass., 1999).
28. For pioneering work in this regard, see Peter Burke, *A Social History of Knowledge: from Gutenberg to Diderot* (London, 2000); Daniel R. Headrick, *When Information Came of Age: technologies of knowledge in the Age of Reason and Revolution, 1700–1850* (Oxford, 2000); C. A. Bayly, *Empire and Information: intelligence gathering and social communication in India, 1780–1870* (Cambridge, 1996).
29. Paul Slack, 'Government and Information in Seventeenth-Century England', *Past and Present* 184 (August 2004), 33–68.
30. Ian Hacking, review of 'A Social History of Truth: civility and science in seventeenth-century England', *Contemporary Sociology* 24: 4 (July 1995), 409.
31. Steven Shapin and Simon Shaffer, *Leviathan and the Air-Pump: Hobbes, Boyle and the experimental life* (Princeton, NJ, 1985); Steven Shapin, *A Social History of Truth: civility and science in seventeenth-century England* (Chicago, Ill., 1994).
32. For an overview, see Peter Dear, *Revolutionizing the Sciences: European knowledge and its ambitions, 1500–1700* (London, 2001).
33. Steven Shapin, *The Scientific Revolution* (Chicago, Ill., 1996), p. 28.
34. Charles Webster, *The Great Instauration: science, medicine and reform 1626–1660* (London, 1975).
35. Alexandre Koyré, *From the Closed World to the Infinite Universe* (Baltimore, Md, 1957). For a recent revision that emphasises the specificity of the history of mathematics in the process, see Michel Blay, *Reasoning with the Infinite: from the closed world to the mathematical universe* (Chicago, Ill., 1998).
36. Quentin Skinner, *Reason and Rhetoric in the Philosophy of Hobbes* (Cambridge, 1996).
37. Ibid., p. 217.
38. Ibid., pp. 5, 294–326.
39. See Jonathan Israel, *Radical Enlightenment: philosophy and the making of modernity 1650–1750* (Oxford, 2001); Margaret C. Jacob, *The Radical Enligthenment: Pantheists, Freemasons and Republicans* (London, 1981).

40. Shapin, *A Social History of Truth.*

41. Ibid., p. 398.

42. Ann Blair and Anthony Grafton, 'Reassessing Humanism and Science', *Journal of the History of Ideas* 53: 4 (1992), 535–40.

43. Shapin, *Social History of Truth*, pp. 355–407.

44. Rob Iliffe, 'Material Doubts: Hooke, artisan culture and the exchange of information in 1670s London', *British Journal for the History of Science* 28: 3 (1995), 285–318.

45. Steve Pincus, '"Coffee politicians does create": coffeehouses and Restoration political culture', *Journal of Modern History* 67 (December 1995), 833; Larry Stewart, 'Other Centres of Calculation, or, Where the Royal Society Didn't Count: coffee houses and natural philosophy in early-modern London', *British Journal for the History of Science* 32: 2 (1999), 133–53.

46. Lisa T. Sarasohn, 'Who was Then the Gentleman?: Samuel Sorbière, Thomas Hobbes and the Royal Society', *History of Science* 42: 2 (2004), 211–32.

47. Davide Panagia, 'Delicate Discriminations: Thomas Hobbes's science of politics', *Polity* 36: 1 (2003), 91–114.

48. Anthony Grafton, *Defenders of the Text: the traditions of scholarship in an age of science, 1450–1800* (Cambridge, Mass., 1991); Anthony Grafton and Nancy Siraisi, eds, *Natural Particulars: nature and the disciplines in Renaissance Europe* (Cambridge, Mass., 1999).

49. Ann Blair, 'Reading Strategies for Coping with Information Overload ca. 1550–1700', *Journal of the History of Ideas* 64: 1 (2003), 11–28.

50. Keith Michael Baker, *Condorcet: from natural philosophy to social mathematics,* (Chicago, Ill., 1975); Charles C. Gillespie, *Science and Polity in France at the End of the Old Regime* (Princeton, NJ, 1981); Ken Alder, *Engineering the Revolution: arms and Enlightenment in France 1763–1815* (Princeton, NJ, 1997).

51. For an illuminating example of how much smoke can be generated from the heat of this scholarship, see the following debate: Charles C. Gillispie and Ken Alder, 'Exchange: engineering the Revolution', *Technology and Culture* 39 (October 1998), 733–54.

52. Sean M. Quinlan, 'Physical and Moral Regeneration after the Terror: medical culture, sensibility and family politics in France, 1794–1804', *Social History* 29: 2 (May 2004), 139–64.

53. Margaret C. Jacob, 'Constructing, Deconstructing and Reconstructing the History of Science', *Journal of British Studies* 36: 4 (October 1997), 459–67.

54. Karen Barad, 'Agential Realism: feminist interventions in understanding scientific practices' in Mario Biagoli, ed., *The Science Studies Reader* (New York and London, 1999), 2; Karen Barad, 'Posthumanist Performativity: toward an understanding of how matter comes to matter', *Signs: Journal of Women in Culture and Society* 28: 3 (Spring 2003), 801–31.

55. See Carolyn Merchant, *The Death of Nature: women, ecology and the scientific revolution* (San Francisco, Calif., 1980); Evelyn Fox Keller, *Reflections of Science and Gender* (New Haven, Conn., 1984); Londoa Schiebinger, *Nature's Body: gender in the making of modern science* (Boston, Mass., 1993).

56. Lorraine Daston, 'The Nature of Nature in Early Modern Europe', *Configurations* 6: 2 (1998), 171.

57. Donna Haraway, *Simians, Cyborgs and Women: the reinvention of nature* (New York, 1991); Bruno Latour, *We Have Never Been Modern*, trans. Catherine Porter (Cambridge, Mass., 1993).

8

medicine, science and intellectual history

deborah madden

We have a historic chance to build a world where all people, no matter where they're born, can have the preventive care, vaccines, and treatments they need to live a healthy life ... There is tremendous untapped potential in the scientific community to address the diseases of the developing world. We've barely scratched the surface of what's possible.

Bill Gates, 16 May 2005[1]

It is at this particular moment, when work is already under way by world leaders to implement the Millennium Development Goals, that we might feel uneasy about the Janus-faced nature of medicine. From its status as a marginal trade in ancient societies, modern orthodox medicine in the West has become a serious professional enterprise commanding considerable resources and research funds.[2] Development in the areas of science and technology has meant that for much of the twentieth century those living in modern industrialised countries could afford to be confident about techniques used to deal efficiently with disease. Public sanitation, vaccination and the curbing of epidemics with antibiotics ensured that medicine more than lived up to its dictionary definition of being a 'science and art concerned with the cure, alleviation, and prevention of disease ... with the restoration and preservation of health'.[3] We might well have anticipated living in a post-disease era, were it not for the arrival of new viruses, such as HIV/AIDS, and a resurgence of old diseases like malaria and tuberculosis.[4] The latest figures for deaths caused by those diseases, as compiled in a recent report from *The Lancet*, showed them to be as plain as '1-2-3': 1 million annual deaths from malaria, almost 2 million from tuberculosis and more than 3 million from HIV/AIDS. Most of these deaths took place in developing countries, where figures for infant mortality are more startling yet: 2.4 million infants die from perinatal causes, 1.9 million from lower respiratory infections and 1.6 from dysentery-related diseases.[5] Out of the eight Millennium Development Goals, three are related to the health of populations living in poverty-stricken areas. By 2015 these goals seek to significantly reduce child and maternal mortality whilst gaining control over epidemics like HIV/AIDS, tuberculosis and malaria.[6]

Set against these high mortality rates, Anne-Emanuelle Birn's *Lancet* report also shows an annual global figure of US$75 billion being spent on medical research, which 'all but ignores the problems of the developing world, as expressed in the so-called 10/90 gap'.[7] Namely, that only 10 per cent of that research related to health actually addresses the problem of disease for 90 per cent of the world's population. More important for our purpose as intellectual historians is the fact that a tiny proportion of this research finds its way into major medical publications or journals.[8] That Birn's article draws attention to this anomaly might be attributed to *The Lancet*'s commitment to international health and a determination, since its origins in 1823, to remain an 'independent and authoritative voice in global medicine'.[9] Yet an implicit bias in contemporary medical journals against diseases connected with poverty reveals a confluence of ideas about disease, politics and ethics, which impacts upon the way medical knowledge is transmitted through texts.

This confluence immediately alerts us to the interdisciplinary nature of medicine and those related but separate subjects of science and technology. Furthermore, the legacy bestowed on writers, critics and academics in all disciplines by the so-called 'linguistic turn' means that any discussion, whether it be a contemporary investigation into global health or an historiographical analysis of patients and practitioners, must seriously engage with the problem of representation. As problematic carriers of 'signs' that are driven by context and change over time, representations of medical knowledge need to be 'unpacked' or deconstructed. Paying attention to the way ideas about disease are transmitted makes it possible to see medicine as a complex form of language activity between 'communities of disputants, authors and respondents'.[10] The discursive, documentary transmission of medicine, far from being a neutral, value-free activity, can be conceptualised as a number of dynamic discourses, sometimes competing, at other times co-existing, which contain certain assumptions, ideologies or paradigms. This historiographical formulation and its proximity to that articulated by the Cambridge School of intellectual historians should be obvious, though it will be specified more precisely, along with a brief survey of how the history of medicine has fared within social and cultural histories. Given the multifaceted nature of the historiography involved, this survey aims to be thematic rather than comprehensive and will lead into a discussion about some of the difficulties involved with an interdisciplinary, contextualised approach. Before this, however, it might prove instructive to investigate contemporary medical and scientific discourses pertaining to global health care, representation and ethics.

i

The interrelated issues of medical representation and ethics are most clearly evident in the area of human genetics, which has often been associated with nineteenth-century eugenics and its agenda of racial purity. A consequence of the Nazi espousal of eugenics to implement its own programme of genocide was that the ethics of scientific and medical research became strictly codified. An abnegation of medical ethics was not exclusive to Germany and doctors working for the American Public Health Service committed ethical misconduct in human research when, between 1932 and 1972, they denied treatment to 399 African-American men for syphilis in Tuskegee, Alabama. Such abuses in ethical and moral standards, combined with societal fears about the creation of a biological underclass, meant that a significant proportion of funding for the Human Genome Project (HGP) was allocated to the study of ethical, legal and social issues (ELSI).[11] HGP, an international enterprise formally begun in October 1990 and completed in 2003, sequenced all of the estimated 20,000–25,000 human genes to make them accessible for further biological study. The ethical component of HGP, however, forms part of an industry making sure that genetic and scientific advances are implemented to benefit rather than discriminate against individuals, communities and populations.

A number of other organisations, such as the United Nations Educational, Scientific and Cultural Organisation (UNESCO) and the World Health Organisation (WHO) also make substantial contributions to the ethics of medical and scientific research, specifically that involving genetics. Its historical composition, combined with the fact that progress in genetics has generally taken place within industrialised nations, has led to anxieties about Western medical research imprinting the norms of its history, tradition and culture onto developing countries. This anxiety even extends to keeping a vigilant eye on organisations like UNESCO and WHO. Recently, there have been calls for a 'Worldwide Insight in Social and Ethical Reasoning' (WISER) to ensure that individuals and communities in poverty stricken or 'lower-resource' countries are represented in a responsible manner and consulted about any developments or changes in medical and scientific research.[12]

For the Bill and Melinda Gates Foundation, an obvious way to improve health in developing countries involves an approach that is based on sound technical, scientific and medical expertise. Endowed with US$26 billion, it has a much larger annual budget than the WHO or, indeed, that of the Global Fund, set up in 2001 to tackle HIV/AIDS, tuberculosis and malaria by Kofi Annan, Secretary-General to the United Nations. The Foundation launched its 'Grand Challenges in Global Health' in January 2003 and offered a 'challenge' to anyone working in the scientific community to develop 'solutions to critical scientific and technological problems that, if solved, could lead to important advances against diseases of the developing world'.[13] The first 14 'Grand Challenges' announced in October 2003 were grouped around seven goals concerned with measuring health, improving nutrition, implementing childhood vaccination, creating new vaccines, treating infectious diseases, controlling vector-borne diseases and curing chronic infection.[14]

All of this would seem very promising were it not for the amount of criticism that has been levelled against the Foundation and its initiatives. Academics from a range of disciplines have cited historical and contemporary studies to show that health-related problems on a national or international scale require a raft of social, political and medical measures. Looking historically at industrialised countries where the decline of mortality took place from the mid-nineteenth century onwards, scholars have pointed to a number of complex factors contributing to this process, including better income distribution, improved diet, developments in public health and medicine, with a rise in literacy and education levels.[15] This has led Birn to conclude that Gates's 'Grand Challenge' calling on the scientific community turns upon 'a narrowly conceived understanding of health as the product of technical interventions divorced from economic, social, and political contexts'. In recent years health care professionals have increasingly rejected a 'hubristic belief' which blithely asserts that they, as exponents of Western medicine, possess 'all of the answers' needed to improve global health.[16] A fully contextualised, integrated and ethical response to health care is one that incorporates a practically orientated compassion, which wishes to find 'meaning in useless suffering'.[17]

Examples here might include work currently being carried out by the Wellcome Trust, which was established in 1936 as an independent charity and now boasts the largest non-governmental source of funding in the UK for biomedical research. Its varied interests in human and animal health also include programmes concerned with malaria control in developing countries. An important part of this sees doctors ascertaining levels of 'informed consent' from patients who agree to blood sampling and drug trials. All medical research, particularly that involved with the ethically loaded issue of informed consent, must receive approval from local and international committees. Doctors working in Kilifi, Kenya, are working to find equivalent terms for 'medical research' as part of the local language (Giriama) because near-equivalents, such as 'discovery' or 'finding out', are unable to distinguish between research and diagnosis. By adapting locally understood concepts, medics are currently developing a specifically localised medical language, which can best describe what happens to blood samples used in their research projects. More efficient channels of communication mean that doctors are hoping to find out what the local community think about the medical research being carried out whilst establishing a bond of trust between doctor and patient.[18]

Diseases relating to poverty, malnutrition and famine cannot simply be remedied by medical technology *per se*, but are linked to powerful structures with attendant issues relating to the redistribution of economic and social resources, such as education and the dissemination of accessible medical expertise. Language barriers present obvious difficulties and here the new information technologies developed by Microsoft are well placed to provide online medical resources to assist with education and clinical practice, though Birn argues that the Gates Foundation should, in fact, urge the international community to refrain from impeding elected governments with redistributive platforms even if debt-servicing were suspended or the profits of multinational or national companies were at stake. He might also support the creation of a fair international system of commodity pricing, inviolable labour standards, and economic incentives for higher wages, increased social spending, and other forms of redistribution, including health care services. This challenge would entail dirty work indeed – not only because much-needed sewerage systems would be built, but because of the enormous threat such an approach poses to existing power structures both locally and globally.[19]

Innovations made by Microsoft during the twentieth century not only established its monolithic status in the world of information technology as master narrative *par excellence*, but gave Bill Gates a firm grip over the very means of global production. Given that it is much easier to apply technical solutions to complicated economic or socio-political problems, it is, perhaps, hardly surprising that the Foundation seeks to remedy the world's ills by using science, technology and medicine as an outlet for its philanthropy. But although the Foundation's aims are very well intentioned, its approach and rhetoric have been heavily criticised for being naïve or outmoded, containing a strong assumption of cultural superiority appended with economic self-interest. The idea that medical technology can remain pure or inoculated from economic, social and political imperatives has been greeted with disdain, scepticism and suspicion, particularly when it is now deemed unthinkable to assume Western techniques might be exported directly to developing countries.

This disease-led strategy is based on what many health care professionals and academics now regard as being an outdated, 'positivistic' medical model, which overlooks the wider context of patients with a range of factors that might be contributing to their illness. A casualty, perhaps, of its own twentieth-century rhetoric, orthodox Western medicine is now less willing to recognise the internal contradictory flux that has always dogged its health. An unshakeable belief in the contingent nature of disease as an unexpected interruption means that medical orthodoxy ignores the presence of illness all around, thus neglecting its responsibility to those who continue to suffer.[20] If community responsibility remains one of its core values, some have argued that the global responsibilities of medicine need to be strengthened in journals to avoid becoming elitist, insular or, more controversially, institutionally racist.[21] A recent survey investigating the under-representation of poverty-related diseases demonstrated that a handful of mainstream medical journals controlled the entire distribution of biomedical news and developments. It concluded that editorial boards had a 'responsibility' to include representative articles identifying important global issues, which could be accessed at a reasonable cost.[22]

Following up on this work, Richard Horton sent evidence of 'institutional racism' to the editorial boards of the *Annals of Internal Medicine*, the *Journal of the American Medical Association*, the *New England Journal of Medicine* and the *British Medical Journal*, but also to the *The Lancet*'s international advisory board. Requesting that editors respond to his findings, their comments formed part of Horton's report, which was subsequently published in *The Lancet*. Judging from the comments made, most of the editors reacted badly to the 'highly emotive and provocative label' of 'institutional racism'. In fact, they fiercely contested the word 'racism', preferring instead to use words like 'elitism', 'bias', 'insularity', 'nationalism' or 'regionalism': the word 'racism' was too politically strident and offensive. Recoiling from actually suggesting that individual editors were, themselves, guilty of racism, Horton nevertheless insisted that the term 'institutional racism' was particularly apposite given its very precise meaning. According to the Commission for Racial Equality in the UK, institutional racism 'occurs when the policies and practices of an organisation result in different outcomes for people from different racial groups'.[23] Words like 'elitism', 'bias' or 'regionalism' obscure and fail to capture the full complexity of this process.

Horton identified the problem as lying with the scientific, medical and public health priorities of Western medicine, which, he says, continue to be represented as being both neutral and normative. Editors seek a global status for their journals, but ignore the actual experiences of those living with illnesses related to poverty. This suggests not only that journals imbibe the attitudes of researchers and practitioners, but that decisions made by editors perpetuate particular biases within medical research. Horton offers two explanations for this: profit and 'impact value'. All journals aim to be profitable, which means that content is aimed at readers and organisations able to afford subscriptions. Although profit can protect editorial freedom, medical research is distorted when commercial imperatives become an overriding concern. In addition to this, research without 'impact value', such as that examining diseases not usually encountered in the West, currently lacks prestige and attracts little interest.[24]

In this sense, as Arthur W. Frank observes, modern medicine remains a 'master text' that needs sick, diseased people, though the wider context of suffering caused by specific, problematic disease and illness cannot be acknowledged.[25] The inherent contradiction contained in this colonising 'master text' becomes more apparent as it struggles to assert itself in a post-colonial context where the experience of illness has been radically redefined.

'Illness', as an individual, social or cultural experience, is something quite distinct from 'disease', which is usually characterised as being a purely physiological condition.[26] Technological advances in modern Western medicine have meant that we expect greater numbers to make a swift and full recovery from disease or illness, though this has produced knowledge of what it means to live in a 'remission society'.[27] While old diseases resurface in tandem with new forms of ever-variant viral infections, we also live in an age dominated by other medical conditions relating to addiction, compulsion, depression and disorder. Certainly in the West we have learned, as Diane Price Herndl wryly observes, to think of ourselves as 'always already subject to the medical gaze', as 'always already' sick. Yet the heroic physician no longer functions to attend a diseased, passive patient. Rather, it is those patients experiencing illness who feel a need to speak and represent themselves.[28] It is now self-evident that any understanding of disease, contemporary or historical, must take into account its contextual relationship to patients and, more crucially, the connection between representations of medical knowledge and ethics. This expanded view of disease is inextricably linked to new types of analysis developed during the 1960s and 1970s, which, by mining and exploiting the interdisciplinary and contextual nature of medicine and science, dramatically altered our view of those subjects.

ii

The historian can number the wills of Tudor or Georgian aristocratic families, but he can never demonstrably 'prove' beyond a shadow of doubt (and even then usually not to the satisfaction of his colleagues) how an idea or complex set of ideas evolved. Like Dido ('moriemur inultae'), the Lovejoys of this world must go to their graves unfulfilled, for such proof does not exist in the history of philosophy and intellectual thought, nor even in the history of theoretical science.[29]

Histories of medicine and science were substantially recast during the 1960s and 1970s, when Max Horkheimer and Theodor Adorno's *Dialektik der Aufklärung* (1944), and its English translation, *Dialectic of Enlightenment* (1973), reached iconic status among radically inclined Marxist and feminist groups eager to challenge the positivistic assumptions of scientific expertise.[30] *Dialektik* was written after its authors, who belonged to the Frankfurt School of philosophers, escaped Nazi persecution by emigrating to America. Contemplating the scale of genocide engineered by the Third Reich, they concluded that the dialectic of European Enlightened thought imprinted its dark side onto the twentieth century, powerfully symbolised in the scientific and technological horrors of Auschwitz. The Enlightenment's uncritical worship of science and reason was directly responsible for this process.

This text grew out of anxieties about a 'crisis of civilisation', which was instigated during the 1930s with the emergence of Nazism and Stalinism, when intellectuals regarded those totalitarian regimes as the portentous crumbling of Enlightened ideals or as a brutal confirmation of its secularising impulse.[31] Horkheimer and Adorno's damning estimation of the Enlightenment was to regard those dearly held principles of reason, freedom, tolerance and knowledge as containing the seeds of their own corruption in the form of nihilism, racial purity and genocidal murder.[32] Their work provided a stark contrast to Ernst Cassirer's *Die Philosophie der Aufklärung* (1932), translated in 1951 as *The Philosophy of Enlightenment*, which had defended the Enlightenment's 'mind' and its forms of thinking, whilst dealing positivism a blow. Cassirer's idealist philosophy turned its attention to topics previously deemed incidental by positivist philosophers, such as culture, religion and aesthetics. By redefining the Enlightenment as 'our contemporary' in this way, Dorinda Outram suggests that *Dialektik der Aufklärung* exposed science and technology as an intersection of 'cultural systems that cannot be readily distinguished from the power structures of society'.[33] The impact made by the Frankfurt School of philosophers during the post-war period significantly determined the way in which medicine and science, as authoritative disciplines, were radically realigned from the 1960s onwards.

While positivist medical and scientific models were being scrutinised by ethically minded clinicians, scientists and health care professionals, the notion of a purely objective truth or realm of knowledge came under attack from a range of academics and scholars. Broadly speaking, thinkers from a number of different countries and intellectual traditions were engaged in parallel projects looking to develop a single paradigm, or interpretative tool, that could be applied across the humanities and social sciences. Under the lead of a French anthropologist, Claude Lévi-Strauss, Structuralism applied methods of order, developed from linguistic theories, to social and cultural phenomena, while Thomas S. Kuhn, an American physicist turned historian-philosopher, marked a relativist reaction to positivism in a text that has since become canonical in most academic disciplines. *The Structure of Scientific Revolutions* (1962), often simply abbreviated to *SSR*, accounted for major transformations in science as 'paradigm shifts', which were the result of social, political and psychological factors.[34]

Kuhn's approach wielded tremendous influence over sociologists, anthropologists and other social scientists casting a critical eye over science, and its relationship to the development, structure and function of human societies. No longer an impartial story of discovery contributing to the progress of objective facts, the historical sociology of scientific knowledge, sometimes referred to as 'SSK', saw science develop in tandem with social and economic forces to become a powerful form of ideology. Inasmuch as it sought to place the social sciences on a par with natural science, SSK was, in the main, less critical of scientific practices, opting instead for sociologically accurate descriptions. Others, however, believed that science and medicine easily mutated to become a tool of coercion serving the ruling class or culturally elite, thus ensuring it was their hegemonic interests that determined medical and scientific development. A study of how this took place in the eighteenth century, often cited by social historians working on the period, was undertaken by Nicholas Jewson who investigated medical knowledge and the patronage system, noting how commercial imperatives, with the needs of the wealthy, shaped orthodox medical practice.[35]

It was in the 1960s, Simon Schaffer observes, that science 'turned dirty'; doyens like George Sarton, the old men from a bygone era, had their vision of the noble science dismantled and turned upside down.[36] John Harley Warner has shown how this period produced an array of scholars wanting to examine historical cases of scientific and medical abuse by way of criticising modern practices. These studies, he says, told 'conspiratorial tales' in which the medical profession was depicted as a 'calculating elite' who used science and medicine to enhance their authority and income.[37] Reflecting on his own experience, Thomas P. Hughes, one of the founding fathers of the Society for the History of Technology, which was formed in 1958, remembers a 'rising hostility' among academics working in the humanities or social sciences and those inhabiting the world of medicine, science and technology. A confrontation between academics inspired by Kuhn's social 'constructionism' and those proponents of scientific 'truth' came to be characterised as the 'science wars', lasting many years and further ignited by cultural, feminist and radical Kuhnian postmodernist studies during the 1980s. This dispute became acrimonious, particularly when, in the case of the latter, all 'totalising' systems and concepts were rejected by way of undercutting the authority of science. Defenders of this discipline fought back by protecting scientific autonomy whilst remaining suspicious about a critique that was rooted in cultural relativism.[38]

As an historian of technology and self-styled 'builder of bridges' between those disciplines, Hughes recalls that many bridges were burnt during the late 1960s, though potent new alliances were also formed. Disenchanted by the hegemony of a science or medicine that was always gendered, frequently misanthropic and certainly militaristic, he sees how historians of medicine, science and technology were stimulated by SSK, and a pressing concern for the social, cultural and economic context of their discipline. This permitted an exploration of convergent interdisciplinary 'themes' between history, science, sociology and cultural studies, rather than a rancorous defence of disciplinary boundaries.[39] Here, the line between political activism and historical practice was sometimes blurred, though studies concerned with insanity or those examining interconnected themes of disease, gender and race amply demonstrated what is now taken for granted. Namely, that medicine and science were ideological 'constructs' based on class, gender, race and geography.[40] Investigating concepts and theories claiming to be purely scientific or medical, scholars allied with the social history of ideas discovered that they were imbued with moral, social and cultural values.

The permutations of these changes to the intellectual landscape were even more profound. As historians of medicine veered away from their parent discipline, drifting inexorably, it seemed, towards social and cultural histories, they found fewer points of contact with the history of science. Increasingly likely to see themselves as historians of 'illness', 'suffering' or 'healing', rather than medicine or disease, they were now less inclined to investigate scientific, technological or medical processes. In part, this cleft grew out of a longstanding distinction between *scientia* and *ars*, theory and practice, with an assumption that although medicine drew on the sciences, in practice it constituted an art.[41] Gradually, as part of a general shift towards the social sciences and humanities, combined with the proliferation of continental philosophy in the writings of thinkers like Michel Foucault and Jürgen Habermas, this division proved more difficult to sustain. In terms of their historiography, histories of science and medicine may have taken diverging routes, but scholars from both fields came to recognise a shared interest in knowledge as construct or 'discursive formation' that was mediated or transmitted to society at large.

Foucault's concept of discourse formed part of a study of language connected to the practice of medicine, criminology and sexology, but spurred historical scholarship well beyond those areas of expertise.[42] Historian, philosopher and polymath, his was an elaborately contextualist methodology that kept its eye fixed on contemporary imperatives. His intellectual and institutional status approached that of divine in France where he was regarded as one of the most dynamic and radical thinkers to emerge from structuralist circles, long before establishing a reputation in the English-speaking world. Self-consciously writing a 'history of the present', it was Foucault's taxonomical, critical scholarship that uncovered exactly how dominant groups in society invoked the authority of scientific and medical discourses to maintain power and control. Working archaeologically, he unearthed specific moments in the history of human understanding – moments that constituted a framework through which we understood our existence or 'fundamental structures of experience'. A change in these structures, defined by Foucault as *epistemes*, involved a paradigmatic shift that necessarily included a transformation in medical perceptions of disease and patients.[43] As part of this shift – in which we, in the present, continue to be deeply implicated – Foucault showed how theories developed by Enlightened 'mad doctors' concerning insanity were heavily laden with political and economic motivations. This contributed to the impure and unethical progress of science, which has since been utilised for technological warfare and annihilation.[44]

Scholars in Britain and the United States took some time to detect Foucault's significance, though, like Jacques Derrida, his international prestige was secured after completing a series of lectures and visiting professorships in American universities. In a review of *Madness and Civilisation: A History of*

Insanity in the Age of Reason (1965), which had been translated from *Folie et déraison. Histoire de la folie à l'age classique* (1961), G. S. Rousseau, now a leading scholar of medical history in his own right, lamented the scant attention this work received outside of France. Noting the importance that this landmark text had for eighteenth-century studies, he praised Foucault for boldly asking questions about the exuberant, contradictory nature of Enlightenment culture – questions hitherto neglected:

> Briefly, the thesis of the *Histoire* is that the seventeenth and eighteenth centuries, particularly in England and France, created the concept *madness* as we have known it in the last two centuries. Madness and the madman were discovered, and ever since 'the face of madness has haunted the imagination of Western man'. Enlightenment thinkers generated the theory to contain and repress what society feared most: the insane were gradually segregated from the (allegedly) sane in a world where madness, not an ever-creeping industrialism or growing technology, was the greatest threat to man. Such creation – according to Foucault – is ultimately an act of repression which has not yet been removed.[45]

It was his search for the rich context of madness that impressed Rousseau most. In writing *Histoire*, he remarks, Foucault had already carved a niche for himself, despite spending much of his time in Parisian cafés criticising existentialism with fellow philosophers: Claude Lévi-Strauss, Roland Barthes and Derrida amongst others. After reading this work no scholar could credibly rely on that well-worn phrase, 'Age of Reason', when Foucault had provided an equally compelling 'Age of Madness'.[46] This was characterised by a period of 'great confinement', beginning with the foundation in 1657 of the Hôpital Général, Paris, which inaugurated an entire network of similarly designated places to confine the poor and mad across Europe.[47] Prior to this, 'madness' had engaged in a free-form dialogue with reason, but was now 'silenced', meaning that the mad were 'shut up' in more ways than one. 'Confinement' was a complex unity in which a 'new sensibility' to poverty and duty of assistance formed part of a work ethic responding to economic problems of unemployment and 'idleness'.

This 'massive phenomenon' was evidenced by an abundance of historical documents showing countless physicians obsessively defining madness as 'disease' – a condition that had for many centuries been explicated in religious terms of Original Sin and Man's Fall from Grace. Medical texts dealing specifically with insanity were virtually absent before 1650 and Foucault noted how the secularising impulse of eighteenth-century medical discourse created a distinction in disease between physics (physiological) and metaphysics (non-physical/spiritual/psychological). Secular medical discourses relating madness to a purely physiological disease underpinned particular political interests at a critical time in French and British history. Designed to protect national power whilst maintaining internal, repressive control, it formed part of an armoury of political and social weapons. A 'clearly articulated perception' of madness radically modified its language with which it could contain dangerously subversive elements in society. The virtue of *Histoire*, but of Foucault's work generally, was that it refused simply to look at medical ideas about disease in isolation from their social, political and cultural context at a time when most histories of medicine tended to treat the symptom of madness *in vacuo*.[48] Foucault examined the aetiology of madness as part of a structurally formed phenomenon which used discourse as an organising principle to show how meaning is constituted through language.

The implications of Foucault's influence are intricate, not least because numbered among his disciples are those who have never read a word of his originating texts. Moreover, those engaged in conscious acts of parricide often repeat the terms of reference already established by this thinker. Scholars owed a debt of gratitude to Foucault's work as emphasis shifted historiographically from tracing the ideological use and abuse of scientific and medical authority towards a more nuanced analysis of what made science and medicine authoritative in the first place.[49] Roy Porter's *Mind-Forg'd Manacles* (1987) covered Foucault's classical age, tracing the changing cultural meanings of madness to show not simply the 'creation' of madness but the invention of its treatment. Porter presented a 'synthesis' elucidating a variety of approaches. Lending the mad an authentic 'voice', he scrutinised the 'relationships between lunacy, literature and the law, between mad people, madhouses and mad-doctors, between attitudes and action, society and psychiatry'. In establishing a 'coherent narrative and integrated analysis' of madness and its treatment from Restoration to Regency, Porter sought to combine the same exacting 'scholarship and chronological sweep' that Basil Clarke's *Mental Disorder in Earlier Britain* (1975) had achieved for the medieval period.[50]

It achieved a great deal more and, by examining the rich density and diversity of reciprocal relationships intersecting in the gaps between medically defined madness and experienced illness, Porter's *Mind-Forg'd Manacles* did for the Enlightenment what Michael MacDonald's *Mystical Bedlam* (1981) had done for the seventeenth century.[51] MacDonald, who had openly acknowledged his source of inspiration as deriving from Keith Thomas's *Religion and the Decline of Magic* (1973), displayed exemplary scholarly skills when contextualising the extensive manuscripts of clergyman-physician, Richard Napier (1559–1634). However, he seemed content to deal explicitly in caricature when assessing the eighteenth century, which he believed had betrayed an earlier age of therapeutic eclecticism with its adoption of cruel medical and scientific treatments for the insane.

Porter corrected some of MacDonald's exaggerated Foucauldian claims about the Enlightenment whilst cauterising the corruption *fons et origo*. He admired the tenacity of Foucault's bold insights, indeed, the 'gloomy reading' of an emerging repressive policy vis-à-vis the insane had a more convincing ring than those histories saluting the 'New Science and Age of Reason as brave new dawns, freeing the mad from being mistakenly identified as demoniac'.[52] The main fault-line in Foucault's revisionist 'age of confinement' was that it simply did not fit facts, not least because of its displacement of human agency from the historical process. Despite this critique, however, methods elaborated by Foucault, along with other philosophers like Habermas, allowed historians like Porter to further develop a 'history of meaning' that could show how knowledge was produced, constructed, disseminated and renegotiated by individuals or groups in various historical settings and cultural contexts.[53]

Jürgen Habermas's *Strukturwandel der Öffentlichkeit* (1962), translated in 1989 as *Structural Transformation of the Public Sphere: An Enquiry into a Category of Bourgeois Society*, suggested ways in which scholars could use his methodology to connect medical and scientific knowledge to a bourgeois 'public sphere'. According to Habermas, the origins of this realm lay in Augustan England, where coffee houses, reading clubs and printing presses spawned an entrepreneurial economy, giving rise to a self-conscious liberal critique. This afforded opportunities for a newly formed enlightened cultural elite to pursue their own interests whilst questioning traditional social and political forces.[54] Habermas's text was important to scholars like Robert Darnton, Roy Porter, Jan Golinski, Simon Schaffer and Steven Shapin, who noted, in their own distinct ways, how medical and scientific knowledge was mediated as the product of a robust, commodified European print culture in 'polite' society.[55] For Porter, an unregulated medical marketplace with a diversity of practitioners offered the full gamut of nostrums and compound medicines to economically enfranchised consumers with purchasing power. By controlling the purse strings, consumers ensured that medical knowledge was tailored to their needs, thereby shaping its development. Darnton, as an early advocate of this methodology in the social history of ideas, used quantitative analysis to disclose literacy rates, readership patterns, publication runs and collections held in personal libraries. This work formed part of an attempt to show the proliferation and consumption of enlightened thinking by 'taking the temperature of various cultural microclimates' in the French provinces.[56]

The influence of Habermas's work over Enlightenment historiographies of science and medicine has since been questioned by Outram who suggests that its use presents problems, particularly when he did not assign a specific place to science in the public sphere. His theory might well have released Enlightenment scholars looking at science from the nightmare of history, that dark view presented by Horkheimer and Adorno, whilst also allowing them to raise questions about the nebulous relationship between cultural history and history of science. Yet Outram is troubled by a description of Enlightenment that sees itself as an 'innocent creation' of reading, writing, conversation and 'amateur science', which, she argues, is one divested from the 'consideration of power'. Historians like Porter and Darnton following Habermas are hard pushed for an explanation as to how or why science should assume any centrality. Nor can the tricky relationship between history of science and cultural history be fully reconciled with the adoption of this theory.[57] Outram is referring here to those specific problems arising out of an attempt to represent histories of science as cultural when, theoretically at least, the conceptual frameworks contain irreducible differences.

Culture designates what is not nature, whilst history of science is concerned with knowledge about the natural world. In terms of their assumptions, cultural historians believe that 'nature' has no independent reality, as such, but is a form of human activity representing a type of knowledge. Traditionally, historians of science have taken the discipline to be concerned with an independently existing universe, though in recent years they have accepted the problems involved with representing this reality.[58] In its approach, cultural history takes something 'strange' or alienated from us, the modern reader, and normalises it within its historical context; a classic model, often cited, is Darnton's *Great Cat Massacre* (1984). Darnton drew heavily on anthropological ideas promulgated by Clifford Geertz, utilising his notion of 'thick description' to explicate fully the meaning of an apparently quirky episode in French history.

History of science, on the other hand, might take something familiar and non-contingent, oxygen, for example, and effectively make this qualitatively strange and distanced from modernity by demonstrating how it is conceptualised in varying ways at different times. However, it was less concerned with historical, contextual differences, believing science and its history could tell us something about the relationship between past and present standards in contemporary scientific practice. During the 1980s and 1990s this mode of interpretation made way for what is usually now referred to as the 'complexity thesis', which rejected holding an 'asymmetrical view' of the past and future by representing historical contexts complete with their complex variables. Thus, what had previously been thought to be the 'conflicting' discourses of science and theology, for example, were now historicised as complicated interactions. David Lindberg, Ronald Numbers and John Hedley Brooke made significant contributions to this historiography and strove to be even more non-Whiggish than their predecessor, Herbert Butterfield.[59]

Steven Shapin, Simon Schaffer, Jan Golinski and Peter Dear, as cultural historians of science, refined further by Dear as 'socio-cultural' historians, created their effect by promoting 'strangeness' and applying it to the subject of science. More familiar aspects of science were represented as contextually bound and investigated through their culture to endow them with 'authentic meaning'. Dear argues that experiment, or experimentalism, held an important place for these scholars because it established an appropriate historical problematic. Experimental activity is visible, thus readily given a sociocultural meaning because of its 'brute practical dimensions'. 'Instrumental' examples of ideas about nature and the 'performance of experimental work' could speak convincingly to those wishing to exempt 'genuine' science from social determination. Conversely, by examining the creation of 'fact' through belief systems, a social activity germinated through culture, these scholars could study groups, communities or, in the case under discussion, 'gentlemen' scientists. In current practice, Dear suggests, differences of approach between these sub-disciplines are ironed out with scholars either ignoring potential problems or raising them by implication only.[60]

As social and cultural historians investigated how medical and scientific ideas are constructed by the culture in which they arose, many observed that this was particularly true of theories about class, gender, race and sexual orientation, which used science to maintain norms deemed socially acceptable. Studies by Porter and T. W. Laqueur showed how masturbation became a target of invective by Georgian medical practitioners who suggested it was a sin and, more importantly, an activity that induced 'wasting conditions' like consumption.[61] By replicating normative values and expectations, Harley Warner suggests that medicine and its scientific foundation guaranteed itself enormous cultural power and control. Key to this process was language and Harold Cook showed the way in which 'professional' medical narratives and rhetorics were constructed in seventeenth-century London to distinguish elite physicians from non-professional lay medics making competing claims to medical knowledge. In so doing, he identified the persuasive nature of professional medicine whilst demonstrating how this came to be accepted and translated into cultural authority.[62]

New Historicist approaches in literary studies further popularised a Foucauldian equality of historical discourse with cultural and social historians hoping to identify even more dynamic contexts in which a variety of individuals actively took part. Scholars searched for meaning well beyond the core of established medical orthodoxy. Historians looked for answers in 'marginal' texts, or discursive practices operating at the fringe of society, which had been neglected by traditional historiographies of medicine. The resulting flurry of scholarly activity devoted to discovering diaries, letters and advice books, which recorded lay perceptions of disease or experiences of illness and self-medication by healers, midwives, and 'quacks', led those inaugurating this 'patient-led' approach to proclaim that the social history of medicine had actually 'come of age'.[63] Exploring the margins could potentially cast light on how discourses at the centre were established whilst offering intrinsic value as a related field. A prominent example was Mary Fissell's *Patients, Power and the Poor in Eighteenth-Century Bristol* (1992), but innumerable works by Porter could be included, the most notable of the genre being *Health for Sale. Quackery in England 1660–1850* (1989), hailed by Laqueur as a genuine 'model' of the social history of medicine.[64] Critics of this approach suggested that historians of medicine needed to be even more true to their pluralistic principles by elucidating the 'multiple meanings' of science rather than seeing it as a 'freighted' term.[65]

The most densely explored area for social historians of medicine tended to be the seventeenth and eighteenth centuries, though Joan Lane's *Social*

History of Medicine 1750–1950 (2001) and Christopher Lawrence's *Medicine in the Making of Modern Britain, 1700–1920* (1994) were the best studies to extend this timeline. Crucially, by identifying the fluidity of medical practice, particularly that carried out during the eighteenth century, a number of scholars exposed the points of contact and difference between medicine, religion and 'superstition'. Much of this interest in religion paralleled, but also informed, new historiographical interpretative methods generally concerned to examine faith and religious belief within the Enlightenment context. Jonathan Barry, Colin Jones and Susan C. Lawrence blurred lines of demarcation between physicians and non-professionals, secular and religious, depicting a context of shared medical knowledge, noting the interaction between politics, piety, charity and philanthropy in the practice of medicine but also, significantly, in the establishment of medical institutions and hospitals.[66] The way in which these accounts presented the manifold historical processes involved when laying the foundations of modern medical institutions demonstrably proved the limitation of relying on a strictly Foucauldian analysis.

By opening out its horizons to incorporate a richer meaning of medical knowledge, methodologies developed by social and cultural histories of science and medicine drew attention to neglected topics. They sought to give 'voice' to previously excluded groups, such as medical lay practitioners, women, or those deemed insane by orthodox physicians. This historiographical innovation was a welcome change, particularly when, as Harley Warner observes, it did much to highlight the glaring inadequacies of traditional histories of science and medicine, such as its hermetically sealed, internal positivist method, which placed emphasis upon ideas formulated by heroic scientists or doctors. During the 1980s a radically inclined version of contextualisation and 'social construction', however, was criticised for privileging its own concerns, de-emphasising longer-established themes equally important when producing a full understanding of the medical and scientific past.

Critics have also detected particular political biases within social histories of medicine generally, which deter scholars from investigating ideas developed by established medical elites. Furthermore, few have been willing to engage with the technical content of medicine and science. This has led critics to argue that social histories have produced accounts which are divested of their scientific and medical specificity.[67] Lack of interest in the history of intellectual thought was linked to a more generalised loss of faith in the belief that rational thinking constituted a universal, objective or positivistic truth in matters of scientific fact or cultural value. Positivistic pride was simply the way in which nineteenth-century scientists and medics strove to differentiate their institutional and professional status from what they regarded as the deeply unimpressive efforts of Georgian science and medicine. Thus ideas produced by scientists, medics or philosophers articulating universalist truths needed to take their place within a much larger context of culturally constructed meaning. That a single mentality, or what A. O. Lovejoy termed 'uniformitarianism', should assume it held the key to unlocking any realm of human experience was deemed morally flawed and intellectually untenable.

Intellectual history itself has often been regarded as having retained its positivistic past, criticised for examining only abstract ideas held by an elite minority. Accusations on this basis have diminished in recent years as work produced by leading scholars in the field showed up the poverty of ideas expressed in this claim. Historians of this discipline have known for some time that they are particularly well placed to consistently renegotiate their traditional practice of historicising and contextualising the intellect and its products.[68] An important caveat here is that the creative interpretation of 'meaningful experience' also shapes our perception of that experience. Its self-conscious search for interpretative strategies to uncover the production and transmission of ideas or texts in context means that the interdisciplinary nature of intellectual history is 'always already' sympathetic to 'dialogical' relationships with other disciplines including science, medicine and technology as well as philosophy, theology, anthropology and linguistics.

A serious and systematic linguistic methodology identifying language as paradigmatic further explicates 'meaningful experience' as rhetorical productions or the constant flow of dynamic, not static, discourses, in which actions are both performed and discussed.[69] Embracing complexity means no definitive perspective on science and medicine will exhaust historical understanding, but intellectual history can resist a radical hermeneutic that denies any reality beyond constructed meanings. Here, courage of conviction has been more than adequately confirmed by robust discussions prompted by its own 'linguistic turn' during the early 1980s. This development owed more to Anglo-American rather than continental philosophy concerned with linguistic theory and, given that J. G. A. Pocock and Quentin Skinner from the Cambridge School of intellectual history had already established their own theory of discourse, or 'rules of the games', this 'turn', though welcome, was somewhat belated.

Historical explanations of specific disease, at particular times in a variety of contexts, can be sensitively attuned to language whilst remaining true to biological realities. We might take note, here, of those studies and histories of public, environmental and global health that are not afraid to confront the material, biological conditions of disease and death whilst also relying on social, political, environmental and economic factors to furnish their accounts. Commenting on how a 'historical reality rooted in existential biological factors' has increasingly re-entered the history of medicine and science, Harley Warner notes that the appearance of HIV/AIDS did much to contribute to this process. No one, he says, can reasonably deny that HIV/AIDS is socially constructed – the product of cultural values, social prejudice and power relationships. Yet the biomedical aspects of this disease cannot be ignored either; to see it simply as a socially determined product would be wilfully mindless and, according to Charles E. Rosenberg, HIV/

AIDS provides a useful nexus for understanding the interaction of biological and social factors.[70] Rosenberg conceptualises disease as being a biological event described by specific verbal constructs containing the intellectual and institutional history of medicine. Its location for public policy, and hence social role, does not preclude an individual identity, which also acts as a site for cultural values to include a structuring element in doctor–patient interactions.[71] Such complexity, with precision and refinement, reminds us that twentieth-century studies conducted by social and cultural historians have been extremely valuable for developing methodological tools to deepen our understanding of how medical and scientific knowledge emerges and changes over time.

iii

New readers at the Bodleian Library in Oxford are required to 'read aloud and sign' a declaration in which they promise, amongst other things, not to 'mark, deface, or injure' any volume or document. Nor should they 'bring into the Library or kindle therein any fire or flame'. The function of scholarship is to preserve and elucidate the record of its tradition. Yet this declaration seems to suggest an anxiety that the very opposite might occur. Faced with the sheer burden of material from and about the past, the historian might easily be tempted to burn rather than interpret the archive. This fear is taboo, held in reserve and never directly referred to, though it gestures towards a central preoccupation: originality of interpretation. To burn the archive would create the possibility of starting again, to forge and invent something distinctive. This provocative thought gives way as the scholar turns, once again, to the text. But here an equally impossible fantasy presents itself; a desire to make historical narrative isomorphic with past experience, to leave no event, thought or feeling unnoticed.

Currently, we search out complexity, which, by turns, 'normalises' the past. We need the protective cushioning of rich context or 'thick description', which by definition threatens to spin forever upon pinpoints of multiple perspective or interpretation. All of this places strenuous demands on the historian, presenting obvious practical difficulties, whilst potentially distorting interpretation or destroying any form of creative, historical imagination. Pursuit of complexity in itself expands contextual meanings but, as David B. Wilson has recently remarked, produces ever-narrower studies void of generalisation.[72] Increased awareness of the great variation of views in different times and places has led some to conclude that ideas about science and medicine are nothing but self-reflections, though this conceptual novelty can, at least, be historicised and added to the archive.

Like John Locke, we have little stomach for grand, overarching, abstract systems, but elaborating principles of differentiation has its own highly wrought theoretical baggage. To borrow an observation made by Suzanne

Marchand when identifying the problems and prospects for intellectual history: texts are often ambiguous, but not *endlessly* so. Part of our task is to work out the 'network of possible variables', which is the 'commonsensical' approach most historians actually put into practice.[73] The fragile nature of an inherently interdisciplinary subject can be easily disturbed, threatened with extinction: subsumed under the weight of social and cultural histories. Yet this threat can also be its very strength, a reminder of what it is that intellectual historians do differently. The apparent interchangeability of 'ideas in context' and 'context of ideas' is more than a grammatical difference in method. The self-limiting nature of its working definition means that intellectual historians know where the boundary of context is located if they want it. Some, of course, may choose to ignore it. As an interdisciplinary approach, intellectual history is the way modern scholars characterise a type of activity that did not need to be justified by early modern scholarship, which sought to create some boundaries or erase others, in a series of conceptual shifts and realignments that could bring entire traditions under consideration.

Examining medical and scientific texts in relation to their own specific vocabularies, can uncover, as John Hedley Brooke has suggested, interesting and surprising relationships between apparently conflicting discourses as they were interacted in the past.[74] An example here can be seen in the medical work undertaken by John Wesley (1703–1791), a controversial public figure who, in many ways, personifies many of the Enlightenment's complexities and contradictions. He continues to be represented in histories of medicine and science as an anti-rational, 'enthusiastic' religious leader who spiritualised healing and peddled 'home-made' folk remedies or 'simples' for his Methodist followers. Emphasis on the populist strains of his medical manual, *Primitive Physic* (1747), obscures a rich variety of biblical, classical and medical sources that were not only informed by the immediate practical needs of its readership, but participated in several significant intellectual debates.[75]

In common with most eighteenth-century medical works, even those written by professional physicians, the preface to *Primitive Physic* pays due acknowledgement to a universe that is structured, though not pre-ordained, by God. Like those physicians, Wesley is careful to treat the discourse of medicine and religion separately by way of maintaining medical credibility and theological integrity. The effect of Wesley's rhetoric, directed in his preface towards 'Faculty' physicians, is suggestive of someone antipathetic to orthodox medicine and its practitioners. Yet all of the remedies listed can be traced back to authoritative sources and judged against the best medical standards of his day. Though not a professionally trained medic, Wesley deployed a detailed knowledge of methods and compound medications used by leading physicians, such as Richard Mead, John Fothergill, William Buchan and George Cheyne, amongst many others. The application of his remedies was underpinned by a deeper belief in the vocation of practical piety, which developed out of a holistic view of nature and healing inspired by Primitive Christianity.

Wesley's use of 'plain speech' in the manual was practically orientated towards his readers. Complex medical knowledge was strictly controlled and mediated through the manual and *Primitive Physic* became a key text that was disseminated by lay preachers. 'Plain speech' was Wesley's way of connecting his medical practice to deeper theological resonances, involving the mediation of God's word. It was also inextricably linked to empirical experimental philosophy, which attempted to free itself from speculative theories. His avowal of 'plain speech' with 'easy and natural method' was a self-conscious move to rhetorically balance the complexity of contemporary medical practice with the simplicity and singularity of an empirical method that derived from God's sovereign power. Simplicity and singularity could defeat the irreducible, multivalent phenomena of disease, which had plagued Man since the Fall, potently symbolised in his adoption of electrical therapy to treat a range of symptoms, including lunacy. Cheap and accessible, electricity was an invisible but materially effective treatment. Wesley's rhetoric of plain speech and simplicity sought to ward off something unholy, beyond human power, whilst rigorously seeking out the latest medical and scientific research to combat physical ills. Safeguarding physical health became, in itself, a spiritual act. For Wesley, physical wellbeing did not depend exclusively on enlightened thinking or spiritual faith but on showing how rationalism and faith could display separate strengths within an overall Christian framework of holism.

Close attention to language means that the careful deployment of knowledge composed in those texts by learned, intellectual elites brings to light a range of hitherto obscured scientific, medical or technical accomplishment. Yet this need not efface the wider social, political, religious or cultural context of this discursive practice or its dissemination to lay practitioners, healers and patients.[76] Conversely, 'rescuing' the sick, diseased, excluded, misunderstood or 'intellectual losers' from the 'enormous condescension of posterity' should not preclude the way in which these groups engaged with or renegotiated definitions of disease formulated by practitioners, the so-called 'winners' in medical history.[77] Indeed, Wesley's patients were not merely the passive recipients of his medical wisdom and remedies set down in *Primitive Physic* were sometimes reinterpreted to fit individual or familial circumstances.

A methodology that can achieve a dialectical unity between these approaches needs to recognise co-existing epistemologies of disease and illness, which were subject to change by individuals wanting to elaborate or respond to other forms of intellectual activity. Drawing on a rich historiographical tradition, intellectual historians can elucidate complex, reciprocal, relationships between science, medicine and society at large whilst remaining in touch with biological reality. This method might direct itself to investigating areas of historical research still under-represented, such as the changeable nature of disease definition and its ethics, in conjunction with looking at the role of language in the phenomenology of illness.[78] As part of globalisation, with the demands and 'challenges' that this will inevitably place on all intellectuals, scholarship might also be enriched and contextualised more meaningfully with an awareness of medical histories and developments in places that have not traditionally fallen under the gaze of Western medicine and science.

notes

1. Quoted in 'Bill Gates calls on world leaders to seize opportunity to improve health', *News-Medical.Net* (published online 16 May 2005) <http://www.news-medical.net/?id=10068>, accessed 4 July 2005.

2. M. and D. J. Weatherall, 'Medicine' in Colin Blakemore and Shelia Jennett, eds, *The Oxford Companion to the Body* (Oxford, 2001), pp. 449–52 (p. 449).

3. *Oxford English Dictionary*, 2nd edn, s.v. 'medicine', <http://www.oed.com>, accessed 4 July 2005.

4. Diane Price Herndl, 'Critical Condition: writing about illness, bodies, culture', *American Literary History* 10:4 (1998), 771–85 (771).

5. Anne-Emanuelle Birn, 'Gates's Grandest Challenge: transcending technology as public health ideology', *The Lancet* (published online 11 March 2005) <http://image.thelancet.com/extras/04art6429web.pdf>, accessed 4 July 2005.

6. Anthony Costello and David Osrin, 'The Case for a New Global Fund for Maternal, Neonatal, and Child Survival', *The Lancet* (published online 23 June 2005) <http://download.thelancet.com/pdfs/journals/0140-6736/PIIS0140673605667037.pdf>, accessed 4 July 2005.

7. Birn, 'Gates's Grandest Challenge'.

8. Ibid.; Richard Horton, 'Medical Journals: evidence of bias against diseases of poverty', *The Lancet* (published online 1 March 2003) <http://www.thelancet.com/journals/lancet/article/PIIS0140673603126657/fulltext>, accessed 4 July 2005.

9. 'About *The Lancet*', *The Lancet* <http://www.thelancet.com/about>, accessed 4 July 2005. Horton argues that whilst *The Lancet* has a higher frequency of articles relating to disease, poverty and global health, it does not quite live up to the ambitious claims made by its editorial board. See Horton, 'Medical Journals'. For comparable statistics, see C. C. Obuaya, 'Reporting of Research and Health Issues Relevant to Resource-Poor Countries in High-impact Medical Journals', *European Science* 28 (2002), 72–7.

10. Quentin Skinner quoted in Richard Macksey, 'Introduction: texts, contexts and the rules of the games', *MLN* 96:5 (1981), pp. v–vii (vi).

11. Arnold Christianson and Bernadette Modell, 'Medical Genetics in Developing Countries', *Annual Review of Genomics and Human Genetics* (published online 19 May 2004) <http://arjournals.annualreviews.org/doi/full/10.1146/annurev.genom.5.061903.175935>, accessed 4 July 2005.

12. Ibid.

13. Quoted in Birn, 'Gates's Grandest Challenge'. This initiative is administered by the Foundation for the National Institutes of Health, the Wellcome Trust and the Canadian Institutes of Health Research.

14. Ibid.

15. T. McKeown, R. Record and R. Turner, 'An Interpretation of the Decline of Mortality in England and Wales During the Twentieth Century', *Population Studies* 29 (1975),

398–9; R. Scofield, D. Reher and A. Bideau, eds, *The Decline of Mortality in Europe* (Oxford, 1991); J. C. Riley, *Rising Life Expectancy: a global history* (New York, 2001); Birn, 'Gates's Grandest Challenge'.

16. Birn, 'Gates's Grandest Challenge'.

17. Arthur W. Frank, *The Wounded Storyteller: body, illness, and ethics* (Chicago, Ill., 1995); Herndl, 'Critical Condition', 774.

18. The Wellcome Trust, 'Malaria Control: Meaning and Research', <http://www. wellcome.ac.uk/en/malaria/MalariaAndControl/mores.html>, accessed 4 July 2005.

19. Birn, 'Gates's Grandest Challenge'.

20. Frank, *The Wounded Storyteller*; Herndl, 'Critical Condition', 773.

21. Horton, 'Medical Journals'.

22. Obuaya, 'Reporting of Research and Health Issues', see note 9 above.

23. Quoted in Horton, 'Medical Journals'.

24. Ibid.

25. Frank, *The Wounded Storyteller*, p. 9.

26. Herndl, 'Critical Condition', 778.

27. Frank, *The Wounded Storyteller*, p. 9.

28. Herndl, 'Critical Condition', 783.

29. G. S. Rousseau, review of *Madness and Civilisation: a history of insanity in the Age of Reason*, *Eighteenth-Century Studies* 4 (1970), 90–5 (95).

30. Max Horkheimer and Theodor Adorno, *Dialektik der Aufklärung* (1944; repr. Frankfurt am Main, 1988); *Dialectic of Enlightenment*, trans. John Cumming (London, 1973); Dorinda Outram, 'The Enlightenment Our Contemporary' in William Clark, Jan Golinski and Simon Schaffer, eds, *The Sciences in Enlightened Europe* (Chicago, Ill., 1999), pp. 32–40 (p. 35).

31. William Clark, Jan Golinski and Simon Schaffer, 'Introduction' in Clark et al., *The Sciences in Enlightened Europe*, pp. 3–31 (p. 4).

32. Alex Callinicos, *Theories and Narratives: reflections on the philosophy of history* (Cambridge, 1995).

33. Outram, 'The Enlightenment Our Contemporary', p. 32.

34. Thomas S. Kuhn, *The Structure of Scientific Revolutions* (1962; repr. Chicago, Ill., 1970).

35. Nicholas Jewson, 'Medical Knowledge and the Patronage System in Eighteenth-Century England', *Sociology* 8 (1974), 369–85.

36. Simon Schaffer, 'What is the History of Science?' in Juliet Gardiner, ed., *What is History Today?* (London, 1993), pp. 73–5 (p. 74).

37. John Harley Warner, 'The History of Science and the Sciences of Medicine', *Osiris* 10 (1995), 164–93 (167, 169).

38. Gary B. Ferngren, 'Introduction', in Ferngren, ed., *Science and Religion: a historical introduction* (Baltimore, Md, and London, 2002), pp. ix–xiv.

39. Thomas P. Hughes, 'Convergent Themes in the History of Science, Medicine and Technology', *Technology and Culture* 22 (1981), 550–8 (551, 552).

40. Harley Warner, 'The History of Science', 167.

41. Harley Warner, 'The History of Science', 164.

42. John E. Toews, 'Intellectual History after the Linguistic Turn: the autonomy of meaning and the irreducibility of experience', *American Historical Review* 92 (1987), 879–907 (890).

43. Michel Foucault, *The Birth of the Clinic: an archaeology of medical perception*, trans. A. M. Sheridan Smith (New York, 1973), 199; trans. of *Naissance de la clinique: une*

archeologie du regard medical (Paris: Presses Universitaires de France, 1963); Michael S. Roth, 'Foucault's "History of the Present"', *History and Theory* 20 (1981), 32–46 (33).

44. Rousseau, review of *Madness and Civilisation*, 93.

45. Ibid., 90, 91.

46. Ibid., 91.

47. Michel Foucault, *Madness and Civilisation: a history of insanity in the Age of Reason*, trans. R. Howard (1967; repr. London, 1999), pp. 45–6; trans. of *Folie et déraison. Histoire de la folie a l'age classique* (Paris, 1961).

48. Ibid., pp. 45–50; Rousseau, review of *Madness and Civilisation*, 94.

49. Harley Warner, 'The History of Science', 169.

50. Roy Porter, *Mind-Forg'd Manacles: a history of madness in England from the Restoration to the Regency* (London, 1987), p. viii.

51. Michael MacDonald, *Mystical Bedlam: madness, anxiety and healing in seventeenth-century England* (Cambridge, 1981).

52. Porter, *Mind-Forg'd Manacles*, p. 7.

53. Peter Dear, 'Cultural History of Science: an overview with reflections', *Science, Technology, and Human Values* 20 (1995), 150–70 (150).

54. Outram, 'The Enlightenment Our Contemporary', 39.

55. Robert Darnton, *The Business of Enlightenment: a publishing history of the 'Encyclopédie', 1775–1800* (Cambridge, Mass., 1979); Roy Porter, *Health For Sale: quackery in England 1660–1850* (Manchester, 1989); Jan Golinski, *Science as Public Culture: chemistry and Enlightenment in Britain, 1760–1820* (Cambridge, 1992); Steven Shapin and Simon Schaffer, *Leviathan and the Air-Pump: Hobbes, Boyle, and the experimental life* (Princeton, 1985).

56. Noted by the editors, William Clark, Jan Golinski and Simon Schaffer, in their introduction to *The Sciences in Enlightened Europe*, pp. 23–4.

57. Outram, 'The Enlightenment Our Contemporary', 39.

58. Dear, 'Cultural History of Science', 153.

59. John Hedley Brooke, *Science and Religion: some historical perspectives* (Cambridge, 1991); David C. Lindberg and Ronald L. Numbers, eds, *God and Nature: historical essays on the encounter between Christianity and science* (Berkeley, Calif., 1986).

60. Dear, 'Cultural History of Science', 153, 161.

61. Roy Porter, 'Consumption: disease of the consumer society?' in John Brewer and Roy Porter, eds, *Consumption and the World of Goods* (London, 1993), pp. 58–81 (p. 67); Thomas W. Laqueur, *Solitary Sex. A cultural history of masturbation* (New York, 2003).

62. Harold J. Cook, *The Decline of the Old Medical Regime in Stuart London* (London, 1986); see Warner, 'The History of Science', 168–9, 172.

63. Claimed by Andrew Wear and quoted by Ludmilla Jordanova, who provided a more sober and scholarly analysis of developments and discussions in 'Has the Social History of Medicine Come of Age?', *The Historical Journal* 36 (1993), 437–49 (437).

64. Thomas W. Laqueur, review of Roy Porter, *Health For Sale: quackery in England, 1660–1850, American Historical Review*, 96:4 (1991), 1195.

65. Harley Warner, 'The History of Science', 193.

66. Jonathan Barry and Colin Jones, eds, *Medicine and Charity Before the Welfare State* (London, 1991); Susan C. Lawrence, *Charitable Knowledge: hospital pupils and practitioners in eighteenth-century London* (Cambridge, 1996).

67. Harley Warner, 'The History of Science', 173.

68. William J. Bowsma, 'Intellectual History in the 1980s: from history of ideas to history of meaning', *Journal of Interdisciplinary History* 12 (1981), 279–92 (279, 280, 283, 288); Toews, 'Intellectual History after the Linguistic Turn', 879–80.

69. J. G. A. Pocock, *Virtue, Commerce and History* (Cambridge, 1985); Bowsma, 'Intellectual History in the 1980s', 290; Toews, 'Intellectual History after the Linguistic Turn', 881.

70. Harley Warner, 'The History of Science', 175; Charles E. Rosenberg, 'Disease and Social Order in America: perceptions and expectations' in Elizabeth Fee and Daniel M. Fox, eds, *AIDS: The Burdens of History* (Berkeley, Calif., 1988), pp. 12–32 (p. 14).

71. Charles E. Rosenberg and Janet Golden, eds, *Framing Disease: studies in cultural history* (New Brunswick, NJ, 1992).

72. David B. Wilson, 'The Historiography of Science and Religion', in Gary B. Ferngren, ed., *Science and Religion: a historical introduction* (Baltimore, Md, and London, 2002), pp. 13–27 (p. 26).

73. Suzanne Marchand, 'Problems and Prospects for Intellectual History', *New German Critique* 65 (1995), 87–96 (88).

74. John Hedley Brooke, *Science and Religion: some historical perspectives* (Cambridge, 1991).

75. A more detailed discussion of Wesley's medical manual in its larger context has been set out in Deborah Madden, *Primitive Physic: John Wesley's 'cheap, safe and natural medicine for health and long life'* (forthcoming). This was extrapolated from 'Pristine Purity: primitivism and practical piety in John Wesley's art of physic', D.Phil. thesis (Oxford University, 2003).

76. Harley Warner, 'The History of Science', 175.

77. See E. P. Thompson's now famous aphorism in *The Making of the English Working Class* (London, 1963), p. 12. In his book review of Roy Porter's *Health For Sale: quackery in England, 1660–1850* (New York, 1989), Laqueur observed that Porter rescued the 'intellectual losers' of Georgian England by taking 'quacks' seriously. See review of *Health For Sale*, 1195.

78. P. Mack, 'Religious Dissenters in Enlightenment England', *History Workshop Journal* 49 (2000), 1–23; Thomas R. Cole, review of *Framing Disease: studies in cultural history*, *Journal of American History* 80 (1993), 660–1 (661).

9

intellectual, social and cultural history: ideas in context

brian cowan

Attention to the context in which ideas were formulated, circulated and received in the past has always been understood as what makes intellectual history *historical* as opposed to being a subset of philosophy or textual criticism.[1] In this sense, then, intellectual history has always been concerned with understanding ideas in their context. But historical 'context' has been defined variously over time and intellectual historians have often used radically different concepts of context, particularly since the 1970s. This chapter surveys the three most prominent ways in which the historical context of ideas have been understood. Put simply, they can be labelled textual contexts, social contexts and cultural contexts. Although all three ways of understanding historical context have thrived at various points in the past and certainly continue to be used today, there is a sense in which general trends in the practice of intellectual history have favoured each of these approaches in turn. At the moment, the cultural history of ideas tends to present itself as the cutting edge of current intellectual practice. This chapter suggests that the present ascendancy of cultural history has developed not just from changing scholarly fashions, but also from a growing apprehension amongst many historians that the older concepts of context utilised by previous intellectual historians were perhaps too restrictive. The cultural history of ideas has flourished in recent years because it offers a self-consciously capacious understanding of historical contexts.

i

Two of the most influential and important intellectual historians of the later twentieth century have surely been Quentin Skinner and John Greville Agard Pocock, both of whom have been relentless contextualisers particularly in the field of the history of political thought. Together, they have made concepts such as 'languages', 'vocabularies', 'discourses' and intellectual traditions absolutely central to the practice of intellectual history. In the work of these two historians and their fellow travellers, intellectual context has been defined almost entirely in terms of understanding new ideas as responses to previously existing ideas. The terms, concepts and habits of thought expressed in these ideas come together to form larger languages, vocabularies or discourses which persist across time and across many different thinkers. They can be traced from text to text, and it is the relationships between these texts that becomes the primary object of the intellectual historian's concern. The ideas studied are almost always found in written texts, and the same is true of their supposed contexts. The result has been an intellectual history in which the study of articulate thought as expressed in written texts has remained at the centre of the field.

When it first began to be expressed in the later 1960s and 1970s, this sort of 'texts as contexts' style of intellectual history dramatically expanded the range of texts that could be studied.[2] While intellectual history had previously tended to concentrate on only a few select texts by the 'great thinkers' of the past, this new focus on ideas in context substantially expanded the number of works which could now be studied. The result was perhaps a quantitative rather than a qualitative shift in the practice of intellectual history. More texts, and more 'thinkers' than ever before were now fair game for study by intellectual historians. The political thought of not only Thomas Hobbes but also Matthew Wren was now worthy of study; the works of John Locke should now be considered in relation to those of Sir Robert Filmer and Mary Astell.[3] Despite this enlargement of the intellectual historian's domain, there remained a notion, more often assumed than fully articulated, that intellectual history was fundamentally the study of great books, even if the total number of possibly great books was now substantially greater than it had been thought to be previously.

This continued reference to, and perhaps even reverence for, past masters characterises the work of both Skinner and Pocock. In the revised version of his programmatic and influential essay on 'Meaning and Understanding in the History of Ideas', Skinner begins by stating that he intends to challenge the notion that 'the task of the historian of ideas is to study and interpret a canon of classic texts'; but he concludes the essay by claiming that 'the classic texts ... can help us to reveal ... not the essential sameness but rather the variety of viable moral assumptions and political commitments'.[4] It has certainly been the case that Skinner's own most distinguished writings have been concerned with explaining the historical significance of a select number of 'classic texts', and indeed texts which had long been considered as classic before Skinner began his attempts to put ideas in their historical context. He has written important and widely cited studies of Machiavelli's *Prince*, Thomas More's *Utopia*, and above all Thomas Hobbes's *Leviathan*, a text which has been a constant source of concern throughout Skinner's career.[5]

Alongside his own influential writings, Skinner has edited an influential monograph series entitled 'Ideas in Context' for Cambridge University Press since 1984 in which close to 100 titles have now been published. This series has from the beginning published books which place various intellectual traditions in their proper context, context being defined by the series description as 'the alternatives available within the contemporary frameworks of ideas and institutions'.[6] Although the majority of the books published in this series have covered the early modern era (1450–1800) and the early modern history of political thought in particular, the series has ranged widely from ancient to modern and even contemporary intellectual history.[7]

Skinner has also edited a sort of companion series for Cambridge University Press, the 'Cambridge Texts in the History of Political Thought', another series of over 100 titles in which student-friendly editions of the 'the most important texts in the history of western political thought' have been provided along with critical introductions and all of the 'contextual' information to allow for historically sensitive classroom discussions of the works. Here too one can detect a tension between a longstanding emphasis on the classic texts of political thought and 'an extensive range of less well-known works'. At one point the series description proclaimed its intention 'to offer an outline of the entire evolution of western political thought', although at present the series aims only to present a 'comprehensive library of political writing'.[8]

The influence of Skinner's own writings alongside those he has edited and sometimes supervised in the form of doctoral theses has been enormous, and especially in his own field of the history of early modern political thought, it can be argued that his approach to studying ideas in context has to a certain degree defined the standard means of doing intellectual history. The intellectual contexts defined by Skinner and his fellow travellers are now very much part of a common set of conceptual apparatuses that are discussed, debated and developed by intellectual historians. The contextual 'languages' of classical (or civic) republicanism, above all, but also Renaissance humanism, civil science, natural law, civil religion and empire, have all figured prominently in the work of this 'Cambridge School' of intellectual history.[9]

The writings of J. G. A. Pocock on the history of political thought have been equally influential and they have fruitfully complemented Skinner's Cambridge-based contextualism with a largely American-based programme of research and graduate instruction that has sought to place early modern political thought particularly within a set of very broad contexts that span geographically from continental Europe to the British Atlantic archipelago to the North American and Pacific outposts of British imperial governance, while the chronological scope has ranged from the high renaissance world of Machiavelli's Florence to the early nineteenth-century British encounter with the Maori peoples of New Zealand.[10]

Like Skinner, Pocock has been equally interested in articulating a more precise, and more historically sensitive, method for practicing the history of political thought and his writings on the topic have further served to elaborate on the concept of political languages, discourses and traditions. For Pocock, to speak of a 'language' in the history of ideas is really to invoke, 'for the most part sub-languages: idioms, rhetorics, ways of talking about politics, distinguishable language games of which each may have its own vocabulary, rules, preconditions and implications, tone and style'.[11] Pocock too has brought to attention a number of different 'sub-languages' of political thinking in his work: classical republicanism especially (and in common with Skinner's efforts) in his *magnum opus*, *The Machiavellian Moment* (1975), but also common law and civil law in his *Ancient Constitution and the Feudal Law* (1957; rev. edn, 1987) and most recently the languages of civilisation and barbarism in *Barbarism and Religion* (1999–2005), his four-volume (to date) exploration of the many mental worlds which informed the writing of Edward Gibbon's *Decline and Fall of the Roman Empire*.

There are of course some significant differences between the intellectual histories of Skinner and Pocock. Skinner has been primarily interested in the ways in which renaissance humanist modes of thinking shaped the 'foundations of modern political thought', and his studies have been largely focused on the sixteenth and seventeenth centuries, while Pocock's work has focused rather on the consequences of certain elements of Renaissance humanist thinking, classical republicanism above all, for later developments in early modern political thought, especially the long eighteenth century that bridges the mid-seventeenth-century British Revolutions and the later eighteenth-century North American and French Revolutions. Methodologically, Skinner's approach owes a heavy debt to the 'speech act theory' of J. L. Austin and John R. Searle, while Pocock has been more comfortable adopting and adapting more structural terms such as 'paradigms' borrowed from Thomas Kuhn or the distinctions between *langue* and *parole* common to French structuralist linguistics.[12]

Nevertheless, the similarities between the two historians' enterprises have on the whole been so striking that both have typically been lumped together, often under the rather imprecise rubric of a 'Cambridge School' of the history of political thought.[13] For both Skinner and Pocock, the contexts in which they have placed the languages or traditions of thought that they study have remained resolutely textual or 'linguistic'. As Pocock has recently put it: 'A great many of the "contexts" in which Skinner and I try to situate political utterances are in fact language contexts, formed by the transmission of "languages" through time.'[14]

There is no doubt that the practice of intellectual history has been enriched tremendously over the course of the past three or more decades by the growth and ultimate dominance of Skinner's and Pocock's detailed efforts to situate past political thinking into their proper 'language contexts'. We know much more now about the nuances and the varied fortunes of habits of political thinking such as civic republicanism, and we are especially well informed about the intellectual milieux out of which strikingly original thinkers such as Niccolò Machiavelli, Thomas Hobbes or James Harrington emerged.[15] We have even added important new works to the 'canon' of texts with which historians of political thought are generally expected to be familiar. James Harrington's reputation was elevated to new heights by Pocock's magisterial efforts to identify him as the key figure in the Anglophone 'Machiavellian moment' and Skinner's 'Texts in the History of Political Thought' series has indeed made reference to the political writings of Mary Astell and Robert Filmer, for example, as convenient as reference to the works of Thomas Hobbes and John Locke has always been.

Yet one cannot help but look back upon the immense labours of this particular contextualist tradition of intellectual history and observe that the concept of 'context' utilised by these historians has been a peculiarly attenuated one. Both Pocock and Skinner have certainly recognised the value of using non-textual contexts, such as the study of socially ingrained *mentalités* or the material histories of printing, bookmaking and text circulation, to situate the work of past thinkers, but in their own work they have largely eschewed the pursuit of these sorts of non-textual contexts in favour of exploring the wide-ranging ideological contexts found almost exclusively amongst the rather articulate writings of intellectuals.[16] Intellectual history practised in this way has therefore remained confined to a closed circuit of contexts: texts are the primary object of study and their meaning is elucidated almost exclusively by reading them with relation to other texts.

ii

At roughly the same time that the 'Cambridge School' of historians of political thought began to revise the canon of classic texts of political theory, a rather different enterprise in placing ideas in historical context was under way primarily amongst French historians and historians of early modern France. Only in retrospect has it seemed to some observers that this approach to the history of the past might well be described as a sort of 'social history of ideas'. Two of the most important practitioners of this approach have been Robert Darnton and Roger Chartier.[17]

While Anglophone intellectual historians prior to the Cambridge School worked primarily within an idealist paradigm set forth by political theorists such as Leo Strauss and Isaiah Berlin, or philosophers such as Ernst Cassirer, all of whom were not primarily historians themselves, Francophone intellectual history developed alongside social and economic history out of the *soi-disant* 'Annales' School of history pioneered by Marc Bloch and Lucien Febvre.[18] Intellectual history in this Annales tradition did not normally go by this name; it was rather considered to be a part of the history of *'mentalités'* or *civilisations*, both of which were crucial terms for this new French history. Although translated only with difficulty as 'mental worlds' or 'cultures', the crucial aspect of these terms was that they sought to draw connections between the social and economic life of the past and the various ways in which people made sense of their existence.[19] The subjects of this history of *mentalités* were rarely articulate texts written by clerics or intelligentsia, but rather feelings, beliefs and superstitions. Two early books by the founding fathers of the Annales School showed the way. Marc Bloch's *The Royal Touch* (1924) examined the rise and fall of a belief in the power of French and English kings to cure scrofula, a skin disease also known as 'the king's evil', over many centuries. Lucien Febvre's *The Problem of Unbelief in the Sixteenth Century: The Religion of Rabelais* (1942) used the question of whether or not François Rabelais was an atheist as a means of exploring the ways in which religious belief were deeply embedded into the structure of Renaissance French society.

With works like these to guide them, Francophone historians could not fail but to develop a very different approach to the mental worlds of the past. And they certainly did. By and large, French intellectual history in the twentieth century developed in conjunction with social history, rather than as a wholly separate field as it did in the Anglophone world. The big questions in French intellectual history rarely centred on texts themselves and when they did, such as with Febvre's use of the works of Rabelais, the texts were more often used as a window into a wider mental world than as the subject of sustained enquiry in and of themselves.

Rather than books about 'Ideas in Context', the French historians Philippe Ariès and Robert Mandrou produced a monographic series published in Paris entitled *'Civilisations et Mentalités'*.[20] Beginning with Louis Chevalier's *Labouring Classes and Dangerous Classes in Paris during the First Half of the Nineteenth Century* (1958), this series has also published titles such as Michel Vovelle's *Baroque Piety and De-Christianization in Eighteenth-Century Provence* (1973) or Mandrou's own *Magistrates and Sorcerers in Seventeenth-Century France: A Psycho-Historical Analysis* (1968). These books were not intellectual histories in the restrictive sense of the term, but rather imaginative studies of the social and mental worlds of the past. The French concept of *mentalité* has deliberately sought to explore the connections between ideas and social life. Works such as these have taken as their province social, religious and magical beliefs, both popular and elite, and they set about to measuring them not through the careful analysis of learned texts alone but also through archival records such as court record or testaments upon death. Vovelle's book, widely recognised as a model of the historical analysis of *mentalités*, relied more on quantitative studies of the numbers of requests for masses found in these testaments than on the careful study of theological treatises about the afterlife.

In this Francophone tradition then, the context in which ideas should be situated is resolutely *social*. True to the continuing influence of Durkheimian sociology and its notions of *représentations collectives*, these scholars have insisted that ideas are not free-floating units, or utterances made by interested actors, they are the products of a social order that represents itself collectively in various expressions, most of which can be measured objectively.[21]

It was not a historian of French extraction who coined the phrase 'social history of ideas', however. The term has rather been associated most commonly with the work of the American historian of eighteenth-century France, Robert Darnton. As an American historian of France who studied at Oxford with the English historian Robert Shackleton, Darnton was well placed to combine the best of a number of different national traditions of intellectual history, and he has done so with a number of influential studies of the ways in which ideas were formed, transmitted and interpreted in pre-revolutionary France.[22] As Darnton himself put it, 'by mixing British empiricism with the French concern for broad-ranged social history it might be possible to develop an original blend of the history of books in America'.[23] Beginning with his first book on *Mesmerism and the End of the Enlightenment in France* (1968), a study of popular attitudes towards science, and continuing apace with the pathbreaking *The Business of Enlightenment* (1979) and his enormously influential collections of essays in books such as *The Literary Underground of the Old Regime* (1982) and *The Great Cat Massacre and Other Episodes in French Cultural History* (1984), Darnton began in the 1970s and 1980s to develop the framework for an understanding of the ways in which ideas circulated in pre-revolutionary France. He did so primarily by 'grubbing' around – a phrase Darnton uses frequently – in the archives.

His particular archival goldmine was the collections of the Société Typographique de Neufchâtel and the roughly 50,000 letters of eighteenth-century correspondence found therein, including letters between agents for the publishing house and its authors, its booksellers, its paper suppliers, its middlemen, and various government officials. This material offered an unusually detailed view of the means by which books were published in old regime Europe. The context in which Darnton situated Enlightenment texts such as the *Encyclopédie* was not primarily other texts, it was the often cut-throat competitive 'booty capitalism' of eighteenth-century publishing.[24] Borrowing self-consciously from the English experience, he called it 'Grub Street'.[25] Grub Street was a real street in London that become renowned in the eighteenth century as a home for impoverished and usually untalented writers, or hacks.[26] By understanding the material constraints, the cultural expectations, and the varied relationships with established authorities experienced by old regime authors, Darnton suggested that we might gain a more complete understanding of both Enlightenment philosophy and even, perhaps, the intellectual origins of the French Revolution.[27]

This was not an intellectual history of the old regime denuded of the classic texts of Enlightenment philosophy – Voltaire, Rousseau and Diderot still figure prominently in Darnton's work – but it is an intellectual history in which the works of these recognised authors are placed alongside those of less well-known writers such as Pidansat de Mairobert, the likely author of a series of scandalous *Anecdotes about Madame Du Barry*, the mistress of King Louis XV. Although it would be an unlikely candidate for inclusion in the 'Cambridge Texts in the History of Political Thought', a translated excerpt from this politically influential eighteenth-century 'bestseller' was included in Darnton's *Forbidden Best-Sellers of Pre-Revolutionary France* (1995). Clearly the criteria for the historical importance of a text are quite different for the social historian of ideas such as Darnton than they are for the intellectual historians of political theory such as Skinner or Pocock.

Criticisms of Darnton's social history of ideas have tended to emphasise its populist character and its supposed neglect of the content of the ideas and conceptual structures articulated by old regime intellectuals, or in other words, for ignoring the ideological contexts that have been the main concern of Cambridge School intellectual historians.[28] Such criticisms are not quite fair, as Darnton's work has been deeply concerned with the various rhetorics of despotism and corruption deployed by critics of old regime institutions, with the different conceptions of 'culture' used in the classic texts of Voltaire and Rousseau, and with the ways in which French cultural identity has been articulated in both written and oral sources.[29] Darnton's tendency to reify the binary distinctions between 'high' and 'low' (or popular and elite) culture often voiced by the subjects of his own research has been slightly more problematic, as perhaps has been his confidence in the revolutionary potential of the 'philosophical' criticisms of the church and the monarchy in the old regime libels he has worked so hard to uncover.[30]

Darnton's social history of ideas has been complemented, and productively criticised, in many ways by his younger colleague Roger Chartier, a French historian who has also been deeply concerned with the relationship between ideas and society in old regime Europe. Beginning with a collaborative study of *Education in France From the Sixteenth to the Eighteenth Century* (1976) with Dominique Julia and Marie-Madeleine Compère, and continuing as co-editor along with Henri-Jean Martin of a four-volume *History of the French Book* (1982–86), Chartier commenced his career with major Annales-inspired 'total histories' of aspects of knowledge production and diffusion in early modern France. In works such as these, 'the social' was the obvious context in which to situate knowledge production, reproduction and diffusion and the best way to document and understand this social context was thought to be through the accumulation of data, primarily quantitative, about the material conditions that made it possible for ideas to circulate. Details about the numbers of schools, the social origins and status of students, the number of books published, the sorts of people who owned them and read them – this is the sort of information that was thought to provide the necessary context for understanding the intellectual life of early modernity. Questions such as 'Were sizable numbers of the general populace among the owners and buyers of books in the sixteenth and seventeenth centuries?' were *questions bien posée* according to this research agenda.[31]

But in the midst of this solidly social history agenda, a new perspective was emerging, particularly in the work of Chartier. In his chapter for the *History of the French Book* on the putatively 'popular' or plebeian books of the so-called *bibliothèque bleu*, a term perhaps best translated as 'cheap print', Chartier took issue with the direct association of cheap print with an easily identifiable popular culture. Most of the texts, he noted, were originally published not for a common, but for an elite readership, and the books only gradually, over the course of more than a century between 1660 and 1780, made their way into a popular peasant culture. He concluded by pronouncing that 'more than in the strictly sociological portrait of their public, then, it is in the *modes of their appropriation* in which the specificity of the "blue" books resides'.[32] This notion of modes of appropriation would later become central to much of Chartier's later contributions to the history of idea diffusion, and it marked a decisive shift of emphasis in his historical writing from prioritising the social as fixed and easily identifiable ('strictly sociological') to a new concept of the social as the product of active creation. It was a turn 'from the social history of culture to a cultural history of the social'.[33]

Chartier's notion of 'culture as appropriation' was certainly influenced by a shift towards recognising agency within social structures in the theoretical works of other French writers such as Michel de Certeau and Pierre Bourdieu.[34] But once expressed, the concept of appropriation allowed historians of various sorts to explore the ways in which various agents – authors, publishers, distributors, and especially readers – took hold of the ideas that were available to them at the time and made these ideas their own. It was an immensely well received notion which Chartier proceeded to develop in a series of influential essays published in the 1980s and 1990s.[35]

Over the course of these two decades, both Darnton and Chartier had become known not primarily as social historians, but as cultural historians. The subtitles of two of Darnton's books were *Episodes in French Cultural History* and *Reflections in Cultural History*, while many of Chartier's essays were translated into English as *The Cultural Uses of Print in Early Modern France* (1987) or simply *Cultural History* (1988), and he wrote a book which explored *The Cultural Origins of the French Revolution* (1991).[36] But neither historian had ever abandoned his interest in the complex interactions between ideas and society. Both wrote important chapters asking the same question, 'Do books make revolutions?'[37] This was the sort of question that begged an answer in terms of a social history of ideas. The difference between the answers offered previous to the 1980s and 1990s and those which followed was that the idea of the 'social' had been seriously reconceptualised; it had taken a linguistic (or cultural) turn of its own.[38] In order to understand the social context in which ideas were developed and diffused, it was clear to historians such as Darnton and Chartier that one must try to understand the various ways in which that context was itself constantly shifting and changing as every social actor 'appropriated' the ideas that circulated around them and tried their best to make their own, often conflicting, sense out of them. The search for a social anchor to the history of ideas would be chimerical if the 'social' itself was a constantly shifting category. Better perhaps, many cultural historians concluded, to concentrate on the ways in which social order was *represented* rather than to engage in a fruitless effort to attach ideas to any concrete social group.[39] In this way, the social history of ideas gradually morphed into a cultural history of ideas.

iii

The current hegemonic status of 'culture' as the reigning context in which to situate idea production, diffusion and reception owes much to the reconceptualisation of the social that took place in the 1980s and 1990s. To a large degree, this movement towards a cultural history has also entailed a certain privileging of the forms in which ideas were conveyed than the content of those ideas themselves. These 'forms' have in practice tended to be understood as the 'norms' of conduct that structured the interactions of intellectuals. In this new cultural history paradigm, past thinkers too become the 'natives' whose foreign customs are now properly the domain of the historical anthropologist. Perhaps the most important intellectual custom for historians of early modern Europe has been 'civility'.[40] The meanings of civility for early modern intellectual life were a major concern for two of the most important and innovative intellectual histories to be published recently, Anne Goldgar's *Impolite Learning* (1995) and Adrian Johns's *The Nature of the Book* (1998).[41]

Goldgar's *Impolite Learning* has perhaps taken the emphasis on the forms in which ideas were presented over the content of those ideas the furthest of any historian thus far. In her account of the culture of scholarly sociability in late Baroque Europe, she questions 'whether scholars' world view was really entirely structured around the subject matter they discussed' and she concludes that in the end, 'when it was necessary to choose between the content of ideas and the formal construction of scholarly society, savants frequently chose the way of moderation, concentration on form'.[42] Although extreme in its explicit denigration of the content of the ideas presented in scholarly writings, Goldgar's emphasis on the importance of understanding the 'socio-cultural context' of these works exemplifies current trends in the cultural history of ideas.[43]

Of course the opposition between the form and the content of ideas is a false dichotomy. *Impolite Learning* tells us much about Baroque scholars' ideas about proper scholarly conduct, even if its author claims not to be interested in the proper 'content' of these scholars' ideas. The notions of civility, politeness and *honnêteté*, all of which figure prominently in Goldgar's book, were of course ideas themselves and these ideas had an important content which is largely explained in the course of this perceptive study. Even the religious beliefs and the attitudes towards toleration of different religious beliefs of her largely Huguenot subjects get substantive treatment in this book.

Goldgar's vocal advocacy of attention to the form in which ideas were conveyed in her work results from her desire to stress that the most important 'context' for understanding the mental worlds of intellectuals in late seventeenth- and early eighteenth-century Western Europe was the prevailing culture of scholarship. The norms, or adherence to the 'communal standards' that comprised the rules according to which membership in the scholarly community was judged, of intellectual life tell us much about what it was that these people who deemed themselves scholars really thought. Her book demonstrates that the notion of a European 'republic of letters' was more than just a notion, it was an actually existing community with standards of conduct according to which its constituent members were judged and evaluated.[44] Rather than demonstrating the priority of form over content in the history of ideas, Goldgar's work in fact demonstrates the necessity of considering the ways in which both the form and the content of ideas work together to form intellectual cultures which have their own histories.[45]

This is a point which has been emphasised time and time again by scholars who have identified themselves as 'historians of the book'.[46] Both Darnton and Chartier certainly incorporated the importance of the history of the book into their understandings of early modern intellectual cultures.[47] But perhaps the most outstanding work in this field in recent years came from a historian of science, Adrian Johns, whose work *The Nature of the Book* shows just how much intellectual labour was required to make the printed text a reliable medium for information dissemination. How could people trust that the information they found in the earliest printed texts was reliable? What were the means by which readers were persuaded that their books, and the knowledge contained in them, were trustworthy? These are the provocative questions posed by Johns in his book.

In addressing these questions of trust and legitimacy in the world of book-making, Johns produces a detailed and persuasive cultural history of ideas. *The Nature of the Book* is fundamentally concerned with the ways in which the representations of knowledge in early modern society. It also emphasises that these representations were always in the process of being made and remade. The work's subtitle – *Print and Knowledge in the Making* – makes it clear that this is a book about the interactions between form and content in the history of ideas about the printed book.

As with Goldgar's book, Johns too places a great emphasis on the concepts of 'civility' that prevailed amongst the authors, the stationers and the regulators of the early modern print trade. Johns defines this civility as 'a congeries of practices and representations used to discriminate propriety from impropriety in everyday life'.[48] By tracking the debates over what was civil or uncivil behaviour in publishing books in the early modern era, and by explaining how a common code of civility came to be accepted in the publishing world, Johns is able to explain how printed knowledge itself was ultimately accepted as not merely legitimate, but fixed and more authoritative than alternative media such as manuscripts or oral communication.

The relationships between texts and their contexts, and between the content and the forms of ideas expressed in *The Nature of the Book* are complex and varied. Johns takes ideas seriously in their own right – political ideologies such as republicanism and legal debates about the importance of custom, precedent and history all figure prominently in his account – but he also takes the social context and the cultural values surrounding those ideas seriously as well. Although this work is clearly indebted to the historical sociology of scientific knowledge, especially the methods outlined by Steven Shapin and Simon Schaffer in *Leviathan and the Air-Pump* (1985), it also demonstrates a solid engagement with the intellectual contexts of its period as well. Johns's book does not assert the primacy of form over content, nor does it try to anchor the rise of print culture in any one social or cultural formation, such as the Stationer's Company, the development of licensing or copyright, or even civility. It offers a cultural history of ideas which recognises that knowledge is always 'in the making' and it shows that a solid understanding of the many contexts in which the processes of knowledge formation takes place is essential to getting the story right.

The Nature of the Book has been criticised for focusing too narrowly on printing practices in seventeenth and eighteenth-century London, a city which was not representative of the European experience as a whole, and a period which is rather too late to capture the truly revolutionary impact of the printing press.[49] But such criticisms perhaps miss the aims of the study, which were not to offer a comprehensive study of printing in early modern Europe nor certainly to relocate the 'printing revolution' in the seventeenth and eighteenth centuries. As with many recent historical monographs, Johns's goals were rather more modest. It is a sort of 'case study' of the ways in which early modern English people thought about the role of print in conveying reliable knowledge and it shows in great detail just how difficult it was for these people to convince themselves that printed information was reliable. When practised in this way, the cultural history of ideas can offer an immensely rich and insightful window into the interactions between ideas and society in the past.

iv

Almost all intellectual historians have recognised the need to contextualise the ideas that they work with. With the recent transformation of the social history of ideas into a cultural history of intellectual life, intellectual historians tend to work at present with two senses of context in which to situate the ideas (and the intellectuals) that they are interested in studying. Cambridge School-style historians continue to produce deeply learned intellectual histories in which the main contexts for their ideas remain the many texts that their subjects were aware of and engaged with mentally. Appreciation of these textual contexts for the historical reading of the works of intellectuals will always be central to the practice of intellectual history, and this style of intellectual history can persist because there is a certain social logic to it as well. Intellectual communities by their very nature tend to be communities defined by their thoughtfulness and particularly by their intense engagement with the thoughts and the writings of other intellectuals, both past and present. The intellectual traditions studied by these historians are real and provide a very important context for understanding what past thinkers thought they were doing in their writings.

Nevertheless, it has also become clear to many other scholars that there are historical contexts which are not textual and can also illuminate the intellectual life of the past. Past thinkers had their own social orders and their own cultural conventions that can be uncovered by the cultural historian of ideas. These thinkers also participated in the wider social order of their times and they also had to engage with the cultural values, prejudices and expectations of their age. Future work on the cultural history of ideas, and the history of intellectual cultures surely has much room for reflection.

The changes in notions of 'context' that intellectual historians have found useful which have been described in this chapter have, to a certain degree, mirrored the changes in other fields of historical enquiry. Political historians have to a large degree abandoned the hope that with enough intensive archival study, they can obtain a complete understanding of past politics and thus write 'definitive studies'.[50] Instead, the history of 'political cultures' tends to dominate. Increasing recognition of the problematic category of 'society' for social historians has also influenced their own turn to an amalgam of social and cultural history.[51] Even economic history has seen a 'culturalist' turn as increasing emphasis is placed by economic historians on cultures of consumption rather than materialist means of production.[52] The dramatic success of these culturalist approaches has led many to suspect that we are all cultural historians now. Perhaps this is so, but the various ways in which culture and its contexts have been construed also suggests that the ascendancy of cultural history will not close down avenues of enquiry, but will rather continue to open up new ones.[53]

notes

1. Given its importance and ubiquity, it is curious that the concept of 'context' has rarely been subjected to more critical analysis in discussions of historical method than it has. But see Bernard Bailyn, *Context in History*, North American Studies Bernard Bailyn Lecture 1 (Melbourne, 1995). I am grateful to Steven Pincus for suggesting this reference.

2. Some of the earliest essays arguing for a contextualist intellectual history may be found in J. G. A. Pocock, *Politics, Language and Time: essays on political thought and history* (New York, 1973) and the essays by Quentin Skinner reprinted in James Tully, ed., *Meaning and Context: Quentin Skinner and his critics* (Princeton, NJ, 1988). Many of the latter have been substantially revised and republished in Quentin Skinner, *Visions of Politics*, 3 Vols, Vol. 1: *Regarding Method* (Cambridge, 2002).

3. See, for example, J. G. A. Pocock, *Virtue, Commerce, and History: essays on political thought and history, chiefly in the eighteenth century* (Cambridge, 1985), pp. 61–8; John Locke, *Two Treatises of Government*, ed. Peter Laslett (Cambridge, 1988); Mary Astell, *Political Writings*, ed. Patricia Springborg (Cambridge, 1996).

4. Skinner, *Visions of Politics*, Vol. 1, pp. 57, 88.

5. Quentin Skinner, *Machiavelli* (Oxford, 1981); Q. Skinner, 'More's Utopia and the Language of Renaissance Humanism' in Anthony Pagden, ed., *The Languages of Political Theory in Early Modern Europe* (Cambridge, 1986), pp. 123–57, revised in *Visions of Politics*, Vol. 2, *Renaissance Virtues*, pp. 213–44; Q. Skinner, *Reason and Rhetoric in the Philosophy of Hobbes* (Cambridge, 1996), and Q. Skinner, *Visions of Politics*, Vol. 3, *Hobbes and Civil Science*.

6. The first volume in the series was Richard Rorty, J. B. Schneewind and Quentin Skinner, eds, *Philosophy in History: essays on the historiography of philosophy* (Cambridge, 1984).

7. One of the most controversial works in the series was Peter Novick, *That Noble Dream: the 'objectivity question' and the American historical profession* (Cambridge, 1988), a book which ended by discussing the contested state of the notion of objectivity in contemporary historical practice.

8. The older series description is quoted from Bolingbroke [Henry St John], *Political Writings*, ed. David Armitage (Cambridge, 1997); the newer description is available at: <http://www.cambridge.org/uk/series/sSeries.asp?code=CTPT&srt=P>, accessed 4 December 2005.

9. Most of these are discussed in Pagden, *Languages of Political Theory*, but see also, for example, Skinner, *Reason and Rhetoric*; Richard Tuck, *Philosophy and Government, 1572–1651* (Cambridge, 1993); J. A. I. Champion, *The Pillars of Priestcraft Shaken: the Church of England and its enemies, 1660–1730* (Cambridge, 1992); and David Armitage, *The Ideological Origins of the British Empire* (Cambridge, 2000).

10. For example, J. G. A. Pocock, *The Machiavellian Moment: Florentine political thought and the Atlantic republican tradition* (Princeton, NJ, 1975); J. G. A. Pocock, *The Discovery of Islands* (Cambridge, 2005); and J. G. A. Pocock, *Barbarism and Religion*, 4 Vols to date (Cambridge, 1999–).

11. Pocock, *Politics, Language and Time*, especially chapters 1, 7–8; J. G. A. Pocock, 'The Concept of a Language and the *Métier d'Historien*: some considerations on practice' in Pagden, *Languages of Political Theory*, pp. 19–38 (p. 21); see also J. G. A. Pocock, 'Texts as Events: reflections on the history of political thought', in Kevin Sharpe and Steven Zwicker, eds, *The Politics of Discourse: the literature and history of seventeenth-century England* (Berkeley, Calif., and London, 1987), pp. 21–34.

12. Tully, *Meaning and Context*; Skinner, *Visions of Politics*, Vol. 1; Pocock, *Politics, Language and Time*; Pocock, 'Concept of a Language'.

13. For example, Melvin Richter, 'Reconstructing the History of Political Languages: Pocock, Skinner, and the *Geschichtliche Grundbegriffe*', *History and Theory*, 29:1 (February 1990), 38–70; Peter L. Janssen, 'Political Thought as Traditionary Action: the critical response to Skinner and Pocock', *History and Theory*, 24:2 (May 1985), 115–46.

14. J. G. A. Pocock, 'Antipodean Historians', *New York Review of Books*, 52: 16 (20 October 2005); cf. Pocock, 'Concept of a Language', p. 22.

15. Skinner, *The Foundations of Modern Political Thought* (2 Vols, Cambridge, 1978), 1: 113–89; Skinner, *Machiavelli*; Skinner, *Reason and Rhetoric*; Skinner, *Visions of Politics*, Vol. 3; Pocock, *Machiavellian Moment* (Princeton, 1975); Pocock, *The Political Works of James Harrington* (Cambridge, 1977).

16. Pocock, 'Concept of a Language', pp. 22, 36–7, and Skinner, 'Reply to my Critics' in Tully, *Meaning and Context*, p. 275.

17. Perhaps the classic statement may be found in Robert Darnton, *The Kiss of Lamourette: reflections in cultural history* (New York, 1990), chapter 11, an essay originally published in 1971. A parallel discussion may be found in Roger Chartier, *Cultural History: between practices and representations*, trans. Lydia Cochrane, (Ithaca, NY, 1988), chapter 1, first published in 1982.

18. The practice of French intellectual history in the twentieth century is ably surveyed in Chartier, *Cultural History*, chapter 1. On Annales history, see Peter Burke, *The French Historical Revolution: the Annales School 1929–1989* (Stanford, Calif., 1990).

19. An important discussion of the concept of *mentalité* is Michel Vovelle, *Ideologies and Mentalities*, trans. Eamon O'Flaherty (Chicago, Ill., 1990).

20. This series, published by the Parisian house Plon, remains active today and is edited by Anthony Rowley and Laurent Theis.

21. See Émile Durkheim, 'Représentations individuelles et représentations collectives' (1898), reprinted in his *Sociologie et Philosophie* (Paris, 1974), and his classic *Elementary Forms of the Religious Life* (1912), trans. Karen E. Fields (New York, 1995).

22. Darnton's career up to the latter part of the 1990s is surveyed in Robert Darnton, 'Two Paths Through the Social History of Ideas' in Haydn T. Mason, ed., *The Darnton Debate: books and revolution in the eighteenth century* (Oxford, 1998), pp. 251–94.

23. Robert Darnton, *The Business of Enlightenment: a publishing history of the Encyclopédie 1775–1800* (Cambridge, Mass., 1979), p. 3.

24. Ibid., p. 4.

25. On his rationale for the term 'Grub Street', see Darnton, 'Two Paths', p. 255.

26. Pat Rogers, *Grub Street: Studies in a Subculture* (London, 1972).

27. The road from Grub Street to the French Revolution was explored in Robert Darnton's *The Literary Underground of the Old Regime* (Cambridge, Mass., 1982); *Edition et Sédition: L'Univers de la Litterature Clandestine au XVIIIe Siècle* (Paris, 1991); and *The Forbidden Best-Sellers of Pre-Revolutionary France* (New York, 1995).

28. Dominick LaCapra, 'Is Everyone a *Mentalité* Case? Transference and the "Culture" Concept', in *History and Criticism* (Ithaca, NY, 1985); Elizabeth Eisenstein, *Grub Street Abroad: aspects of the French cosmopolitan press from the age of Louis XIV to the French Revolution* (Oxford, 1992); William Palmer, 'Exploring the Diffusion of Enlightened Ideas in Prerevolutionary France', *The Eighteenth Century: Theory and Interpretation* 26 (1985): 63–72; Daniel Gordon, *Citizens Without Sovereignty: equality and sociability in French thought, 1670–1789* (Princeton, NJ, 1994), pp. 137–9; Daniel Gordon, 'Beyond the Social History of Ideas: Morellet and the Enlightenment' in

Jeffrey Merrick and Dorothy Medlin, eds, *André Morellet (1727–1819) In the Republic of Letters and the French Revolution* (New York, 1995), especially pp. 45–6; and the many essays in *The Darnton Debate*, ed. Haydn T. Mason, in which Darnton offers a lengthy reply to these criticisms.

29. For the last, see Robert Darnton, *The Great Cat Massacre and Other Episodes in French Cultural History* (New York, 1984), chapter 1, and the criticism in Chartier, *Cultural History*, chapter 4; Darnton's reply in *The Kiss of Lamourette*, chapter 15; and the further observations in Dominick LaCapra, 'Chartier, Darnton, and the Great Symbol Massacre', *Journal of Modern History* 60 (1988), 95–112; and James Fernandez, 'Historians Tell Tales: of Cartesian cats and Gallic cockfights', *Journal of Modern History* 60 (1988), 113–27.

30. See the criticisms of Roger Chartier, *Forms and Meanings: texts, performances, and audiences from Codex to computer* (Philadelphia, Penn., 1995), chapter 4; and *The Cultural Origins of the French Revolution*, trans. Lydia Cochrane (Durham, NC, 1991), chapter 4.

31. Roger Chartier, *The Cultural Uses of Print in Early Modern France*, trans. Lydia Cochrane (Princeton, NJ, 1987), p. 146; a translation of an essay originally published in 1982 for the *Histoire de l'édition française*.

32. Chartier, *The Cultural Uses of Print*, p. 263, emphasis added; a translation of an essay originally published in 1984 for the *Histoire de l'édition française*.

33. Roger Chartier, 'The World as Representation' (1989) in Jacques Revel and Lynn Hunt, eds, *Histories: French constructions of the past*, trans. Arthur Goldhammer et al. (New York, 1995), p. 549. See also Burke, *The French Historical Revolution*, p. 84.

34. See Roger Chartier, *On the Edge of the Cliff: history, language, and practices*, trans. Lydia Cochrane (Baltimore, Md, 1997) especially chapter 3; Chartier, 'The World as Representation', pp. 544–58; and see Michel de Certeau, *The Practice of Everyday Life*, trans. Steven Rendall (Berkeley and Los Angeles, Calif., 1984), especially chapter 12, for an influential discussion of readers as appropriative agents.

35. Roger Chartier, 'Culture as Appropriation: popular cultural uses in early modern France' in Steven L. Kaplan, ed., *Understanding Popular Culture: Europe from the Middle Ages to the nineteenth century* (Berlin, 1984), pp. 230–53; Chartier, *Cultural History*; Roger Chartier, 'Texts, Printing, Readings' in *The New Cultural History*, ed. Lynn Hunt (Berkeley and Los Angeles, Calif., 1989), pp. 154–75; Chartier, *Forms and Meanings*, especially chapter 4.

36. Darnton's books were *The Great Cat Massacre* and *The Kiss of Lamourette*.

37. Chartier, *The Cultural Origins of the French Revolution*, chapter 4; Darnton, *The Forbidden Best-Sellers of Pre-Revolutionary France*, section III.

38. This *soi-disant* linguistic turn was evident in just about every field of historical enquiry at the time, but the debates were particularly lively in the field of social history, precisely because so much had been invested in the theoretical stability of the 'social' by the many social historians. In British social history, the catalyst for these debates was surely Gareth Stedman Jones, *Languages of Class: Studies in English working class history, 1832–1982* (Cambridge, 1983), the implications of which became all the more clear after a decade of reflection. See David Mayfield and Susan Thorne, 'Social History and its Discontents: Gareth Stedman Jones and the politics of language', *Social History* 17:2 (1992), 165–88, and the ensuing debates: Jon Lawrence and Miles Taylor, 'The Poverty of Protest: Gareth Stedman Jones and the politics of language – a reply', *Social History* 18 (1993), 1–15; Patrick Joyce, 'The Imaginary Discontents of Social History: a note of response to Mayfield and Thorne, and Lawrence and Taylor', *Social History* 18 (1993), 81–5; David Mayfield and Susan

Thorne, 'Reply to "The Poverty of Protest" and "The Imaginary Discontents"', *Social History* 18 (1993), 219–33; James Vernon, 'Who's Afraid of the "Linguistic Turn"?: the politics of social history and its discontents', *Social History* 19 (1994), 81–97; and Geoff Eley, 'Is All the World a Text? From social history to the history of society two decades later', in Terence McDonald, ed., *The Historic Turn in the Human Sciences* (Ann Arbor, Mich., 1996), pp. 193–244.

39. Chartier, 'The World as Representation'.

40. See, for example, the essays on this theme collected in the *Transactions of the Royal Historical Society* 12 (2002).

41. As Goldgar notes in a note to her book, there has been something of 'a *zeitgeist* in historical thinking about the importance of civility to academia' in recent years: Anne Goldgar, *Impolite Learning: conduct and community in the republic of letters, 1680–1750* (New Haven, Conn., 1995), p. 253 n.28. A major influence here has been the work of historians of science, for whom the sociology of knowledge has always been central. The seminal work here is Steven Shapin and Simon Schaffer, *Leviathan and the Air-Pump: Hobbes, Boyle and the Experimental Life* (Princeton, NJ, 1985); but see also Mario Biagioli, *Galileo, Courtier: the practice of science in the culture of absolutism* (Chicago, Ill., 1993); Paula Findlen, *Possessing Nature: museums, collecting and scientific culture in early modern Italy* (Berkeley and Los Angeles, Calif., 1994); and Steven Shapin, *A Social History of Truth: civility and science in seventeenth-century England* (Chicago, Ill., 1994).

42. Goldgar, *Impolite Learning*, pp. 5, 7.

43. Ibid., p. 5. The critical reaction to this book has been lukewarm, mostly due to its extreme refusal 'to discuss the intellectual issues' at stake in the texts that intellectuals spent most of their time working on. See, for example, Robert Iliffe, review in *Economic History Review* 50:2 (May 1997), 395–6 (p. 396); compare Roger Chartier, 'When Scholars were Civil', *Times Literary Supplement* 4872 (16 August 1996), 12.

44. Goldgar, *Impolite Learning*, pp. 3, 116, and see chapter 3 *passim*.

45. Apparently effortless demonstrations of this may be found in the numerous recent studies of Anthony Grafton, for example: *Bring Out Your Dead: the past as revelation* (Cambridge, Mass., 2004); *The Footnote: a curious history* (Cambridge, Mass., 1999), and *Forgers and Critics: creativity and duplicity in Western scholarship* (Princeton, NJ, 1990).

46. A true pioneer in this field was Donald F. McKenzie, a scholar whose work demonstrated the importance of bibliography for historical understanding and vice versa. See D. F. McKenzie, *Bibliography and the Sociology of Texts* (Cambridge, 1999), and *Making Meaning: 'Printers of the Mind' and other essays*, ed. Peter D. McDonald and Michael F. Suarez (Amherst, NY, and Boston, Mass., 2002).

47. Darnton, *Kiss of Lamourette*, chapter 7; Chartier, *On the Edge of the Cliff*, chapter 6.

48. Adrian Johns, *The Nature of the Book: print and knowledge in the making* (Chicago, Ill., 1998), p. 632.

49. Elizabeth Eisenstein, 'An Unacknowledged Revolution Revisited', *American Historical Review* 107:1 (February 2002), 87–105; see also the reply by Adrian Johns, 'How to Acknowledge a Revolution', and Eisenstein's response, in ibid., 87–128, as well as Elizabeth Eisenstein, *The Printing Revolution in Early Modern Europe* (2nd edn, Cambridge, 2005), especially pp. 313–58.

50. Ronald Hutton, *Debates in Stuart History* (Basingstoke, 2004) discusses many of these changes in seventeenth-century political history.

51. Witness the founding of a new journal *Cultural and Social History* in 2004, and the commencement of a new monograph series 'Cambridge Social and Cultural Histories' by Cambridge University Press in the same year.

52. To take one individual example from a massive literature, see the shift in emphasis from Maxine Berg, *The Age of Manufactures, 1700–1820: industry, innovation, and work in Britain* (Oxford, 1985), to Maxine Berg, *Luxury and Pleasure in Eighteenth-Century Britain* (Oxford, 2005).

53. For a recent critique of certain aspects of cultural historical practices, see Peter Mandler, 'The Problem With Cultural History', *Cultural and Social History* 1 (2004), 94–117, and the responses from Carla Hesse, Colin Jones and Carol Watts, in ibid., 1 (2004), 201–24, as well as Mandler's reply to them in ibid., 1 (2004), 326–32. The tone of this debate is much more moderate, and the participants perhaps more optimistic about the continued viability of cultural history, than had been the case in similar debates about the 'linguistic turn' in historical studies in previous decades.

10
gender and intellectual history

rachel foxley

The study of the gendered aspects of intellectual history does not enjoy the dubious blessing of being a distinct and well-developed discipline, or even sub-discipline. Perhaps this is as it should be: there are so many aspects of intellectual history which demand gendered attention that it would be unhelpful and reductive to try to unify the field. The range of the work currently being done which might in some way be categorised as gendered intellectual history, or the intellectual history of gender, is staggering. Yet I think there is something to gain here from considering some issues in this range of work under one heading; and this chapter may also point up the fact that, even within this great range of work, there are areas where there is still much to be done.

This chapter offers the view of a historian rather than a philosopher. Obviously the history of ideas is a subject area which cannot be neatly parcelled out between disciplines, and that is its fascination. However, systematic treatments of the place of gender in the history of ideas by feminist philosophers, rather than by historians, reveal a set of concerns which may not be quite the same as those of historians.[1] In this chapter I will attempt to ask how gender impacts on the questions historians tend to ask of past ideas, rather than on the questions philosophers tend to ask. Feminist philosophers have suggested, however, that the type of question asked may itself be affected by the need for a gendered approach, and this may also be true of the questions historians need to ask. I will therefore consider not only the place which gender could take in intellectual history as it is usually done, but the possibility that attention to gender may challenge some of our beliefs about the 'right' way to do intellectual history. The interpretation of ideas and texts from the past in gendered terms may bring up different issues from those raised by other interpretations of texts from the past – but equally it may bring up issues which all intellectual historians would do well to consider.

i

Feminist scholarship over the last few decades has brought into focus the fact that 'gender', and even 'sex' itself, has a history. The natures ascribed to women and men, and the roles prescribed for them, are historically and culturally specific. Even at the level of biology itself – often seen as the natural bedrock of 'sex' on which the cultural roles of 'gender' are then built – one might (given the evidence) write a history which involved considerable change (for example, in the age of menarche), and aspects of biology which may seem basic and foundational to a given society can be perceived totally differently by another.[2] Deconstructive approaches have also challenged the security of the category of (biological) sex, implying that any history of 'women' is a history of a shifting and artificial category.[3] Depending on how broadly one defines intellectual history, much of the work that has been done in gender history can be brought within it: an intellectual history extended from a culture's 'high' philosophical texts to its everyday discourse and even its (often unarticulated) cultural assumptions would have plenty of scope for considering shifting assumptions about women and men, and specific uses of gendered discourses. The history of people's scientific, social, religious and political understandings of the natures and roles of the sexes could well be described as intellectual history.[4] And arguably no history of the lives of women as women, or of men as men, would be an adequate one without reference to the beliefs about gender that shaped those lives.

Here, then, is one large area in which gender and intellectual history come together. Margaret Sommerville, in her book *Sex and Subjection: attitudes to women in early-modern society*, offers a good example of the richness and precision that is possible in this field. The patterns of thought which she documents are generally those of the intellectual elite – jurists and divines as well as philosophers. Her history of their thinking makes close use of the sources of their thought – classical, scriptural, patristic, and the later European traditions drawing on these – and she is particularly good at showing the (relatively rare) points at which writers within her period departed from the views which had become standard in these traditions, and their specific motivations for doing so.[5] This is undeniably intellectual history, and of a relatively traditional type. But where she alludes to popular views and practices which do not coincide with those she describes,[6] we glimpse another possible intellectual history of gender, which would have to be constructed from far less articulate sources, but which does, after all, cover beliefs about all the same things which elite sources wrote about. At this point, almost entirely due to the nature of the available sources, intellectual history begins to blur into cultural and social histories of gender.[7]

The history of women, men, and gender roles thus contains an important element of intellectual history. But I want to argue that the presence of gender in intellectual history is much more extensive than this alone might suggest. This is because gender is often present in texts and thought in extended, metaphorical ways. This is so pervasive a phenomenon that any area of intellectual history – even, say, such a supposedly gender-free subject as the history of mathematical concepts – may well, on examination, turn out to have a gendered dimension.

Essentially, 'gender' is differentiated from 'sex' by being a cultural construct: as historians of gender, we assume that gendered notions have no necessary connection with biological sex. Femininity has no necessary connection with being female, nor masculinity with being male. In a sense, then, gender itself is metaphorical. Femaleness is taken to have ramifications which it does not necessarily have: a particular set of attributes which a particular society at a given time views as femininity. This extension of femaleness into femininity can be taken further: femininity can be attributed to people, things or phenomena which are not female. This extension of the meanings not only of maleness and femaleness, but of masculinity and femininity, means that gender gets into everything. Even texts which have nothing to say about the sexed roles of women and men may use gendered language in describing apparently ungendered phenomena or arguing for apparently ungendered conclusions.[8] This means that the history of human thinking is full of gender, and that intellectual historians need to pay close attention to the operations of gender in the texts and systems of thought that they study.

As an illustration of what I mean by the metaphoricity of gender, consider the 'Pythagorean Table', a set of contrasted qualities originating in the thought of Pythagoras and his followers in the sixth century BC, and known to Plato and Aristotle:[9]

Limit	Unlimited
Odd	Even
Right	Left
Male	Female
Resting	Moving
Straight	Curved
Light	Darkness
Good	Bad
Square	Oblong

The significance of these opposed terms for the Pythagoreans was rooted in their specific philosophical system, and some of these pairings are rather alien to our way of thinking. What they illustrate well is that maleness and femaleness can come to have associations well beyond the obvious qualities of men and women. 'Male' and 'female' are seen as abstract qualities, comparable with odd/even, straight/curved, right/left. On the face of it, this is surprising, as the other qualities on the list seem to be far more abstract: predicable of a far wider range of objects in the world, and only to be grasped as qualities in their own right by a serious effort of generalisation, comparison and abstraction. Not only are 'female' and 'male' treated as abstract qualities whose opposition is an important fact about the structure of reality, but they are aligned, in their columns, in a way which makes a valuation of them clear: female in the 'bad' column, and male in the 'good'. These columns also associate each of them directly with a range of highly abstract qualities.

What does the Pythagorean Table tell us about how we could approach gendered intellectual history? The mixture of pairings whose significance seems culturally remote (resting/moving, square/oblong; and would we now value 'odd' above 'even'?) with those which we might still recognise (light/darkness aligned with good/bad) should remind us how much care is necessary in reconstructing the (gendered) value systems of the past. Some values can be remarkably stubborn and culturally widespread (notably, the valuation of right over left and male over female), but they can become associated with others which are not only specific to a particular culture at one point in time, but even to a particular philosophical worldview. To interpret the Pythagorean Table one would need a good understanding of the significance in Pythagorean thought of 'limit' (what is definite) and 'unlimited' (the indefinite), ideas which were bound up with the mathematical nature of parts of Pythagorean philosophy; the precise significance – rather than simply the negative valuation – of aligning 'female' with 'unlimited' might then become clear.

Even more care would then be needed in reading back femaleness into other references to the 'unlimited', either in other Pythagorean material or in other Greek authors. Scholars who have used the Pythagorean Table to interpret aspects of the thought of Plato and Aristotle have had to explain how Pythagorean 'limit' and 'unlimited' relate to 'form' and 'matter' in these later Greek philosophers, and have substantiated the male/female associations of form and matter from, among other sources, views on reproduction found in Aristotle and elsewhere, where the male parent provides the 'form' of the offspring, and the female only the 'matter', the menses which do not flow during pregnancy.[10] Teasing out such networks of value within an author's works or a culture's thought is always difficult, and in most cases there is no helpful table of values to use as a 'key'. But it is frequently possible to see that gender is being extended metaphorically into other areas of thinking, and to try and reconstruct the significance of this. The presence of such networks of gendered associations in texts from the past can open up fertile new avenues for interpretation.

An influential example of such an approach is the work of Genevieve Lloyd, which draws out the symbolic 'maleness' of reason in the Western philosophical tradition.[11] Although her work came relatively early in the development of feminist history of philosophy, and has been critiqued from various perspectives, it offers a very suggestive overall thesis and is careful in its delineation of the gendered implications of successive philosophical accounts of reason. Her evidence for a gendered notion of reason spans from the male and female associations of form and matter, mentioned above, to the metaphors of marriage which Bacon employed in describing the proper relationship of the enquirer to Nature. However, while there are enough continuities in cultural beliefs about femaleness – for example, its association with corporeality – to implicitly gender successive theories of reason male by contrast, Lloyd refrains from suggesting that the theories which follow each other are gendered in the same ways; as views of form, matter, knowledge and the passions are remade, so are theories of reason, and with them the gendered nature of those theories – even if reason is always gendered male.

Lloyd's work and other work of this kind raises the issue of how deeply these metaphors inhabit the thinking they are found in. Can they simply be peeled off from the surface expression of thoughts, leaving a respectably ungendered theory beneath? I would want to offer two significant cautions here. Firstly, the 'expression' of thought is all we have. We cannot access the pure, unexpressed thought of historical figures – even if we optimistically believe that such a thing did exist in their own minds before they wrote. So to start picking away parts of our only evidence for their thought is, in a sense, to deface the sources we have, and potentially to subtract from the clues we have to their understanding. After all, these authors used these metaphors *in order to* express their meaning. Even if they no longer function as well as they would have done for their original audiences – a sexist metaphor perhaps has a shorter lifespan than, say, an epistemological theory – we still, as historians, need to consider how the metaphors were supposed to help that first audience. Secondly, to rehabilitate the theory of a past thinker by reinterpreting it in a way which dispenses with its gendered or sexist elements may be a valid operation, but it is not the same as interpreting it in its historical form. If we are approaching these texts as historians, these parts of their historical meaning need to be given their full weight. The metaphorical presence of gender in many kinds of intellectual history cannot be so easily stripped away.[12]

Having demonstrated a range of ways, from the most literal to the most metaphorical, in which gender can be at issue in the interpretation of texts and ideas from the past, I would like to illustrate this in the particular case of the history of political thought.[13]

The history of political thought offers fertile ground for the historian of gendered thinking. Within the political writings of the Western tradition, gender is present in a whole range of incarnations, from the most concrete issues of women's (and men's) relation to political and social life, through to the most abstract uses of gendered language in political arguments that are not overtly concerned with the roles appropriate to women and men.

This may surprise some, as politics is often seen as the sphere of Western life which has been most exclusive of women almost throughout its history. However, women have often been present in politics – even if they were exceptional women – and their influence there has led to the discussion of women within political theory. Dynastic monarchical systems tended to come up periodically against the issue of female rule,[14] or at least female influence on male rulers. Dynasties, being families, inevitably involve women, even where a Salic law keeps them out of the line of succession itself. A political debate about a particular woman may bring into play exactly the webs of metaphorical associations of gender which were at issue in the Pythagorean Table. People thinking about the role of Charles I's Catholic wife, Henrietta Maria, might mobilise political stereotypes of gender as well as of Catholicism.[15]

Even at times and places where women have been more thoroughly excluded from the practice of politics, they are extremely present in political thought. The exclusion of women from political practice requires a lot of work at the level of theory, and authors struggling to define the nature and boundaries of the political have often given much attention to women and the feminine, as well as to men and the masculine. Aristotle's *Politics*, a foundational text in the Western tradition (and one by an author often seen as the father of Western misogyny), has a lot of time for women. By the second page, Aristotle is discussing the 'distinction between the female and the slave',[16] apparently unaware that in doing so he is drifting, even before he has started, away from the proper subject matter of politics. Aristotle is attempting to delineate the scope and purpose of political activity within society, and a dissection of all the elements of the polis (city-state) and how they relate to its activity as a polis – its properly political life – is crucial to his account. A close reading of his account reveals that women play a fascinating and troublesome part in it. Aristotle's concern to define the political leads him to discuss slavery as the antithesis of political modes of relationship. But, as I mentioned, women and slaves are to be distinguished from each other, and thus even though women are not politically active citizens in the polis, they do have to be shown relating to male citizens in a way which is not slavish: they are not simply to be ruled with no concern for their own interests. This means that when Aristotle sets up parallels between types of domestic relationship and types of rule within states, women are far from the bottom of the heap: they are ruled not by the despotic or royal rule suitable for slaves and children, respectively, but by either 'political' or 'aristocratic' rule – both of which recognise a significant degree of merit in the person ruled.[17]

The presence of gender in political theory goes beyond the literal concern about the status of women. The realities of gender roles and family life could provide fertile metaphors for political events. Marriage became drawn into debates about political life in the seventeenth century as the idea of a relationship which was contractual and based in consent, but nonetheless irreversible, became a useful way for royalists to conceptualise the contract entered into by people and rulers. This meant that those opposed to royalist views in politics were drawn into debates about the marital analogy.[18] Once such parallels begin to be argued over, they can pull both ways – on the view which is to be taken of political life, but also on the view which is to be taken of marriage and divorce.

At an even more metaphorical level, gender can become a part of arguments in the almost abstract ways that the Pythagorean Table has prepared us for. Modes of political behaviour can be characterised as masculine or feminine; and as I will suggest later, in the example which closes this chapter, whole traditions can become associated with femininity or masculinity. For Milton it was quite comprehensible to describe a whole state as 'manly' because of its commitment to republicanism.

ii

Having given some examples of the ways in which gender is present in the sources studied by intellectual historians, it is now necessary to think about how it has been and should be studied, and why. The scholarly work that has been done exemplifies numerous approaches, and the work being done now has no unifying methodology or set of concerns – particularly as its authors come from a variety of disciplines. One contour that can clearly be seen, however, is a change over time in the way feminist scholars or scholars of gender have read these texts from the past. I will start by outlining some of these changes; move on to consider the multitude of modes in which gendered intellectual history is now being done, and then consider some of the issues of methodology that arise. Again, the majority of my examples will come from the history of political thought.

The first surveys of the place of women in the Western tradition of political thought began to appear in the 1970s and 1980s, as part of the campaign of politically active feminists within academia to challenge and transform disciplines which they saw as male-dominated and male-oriented. These were pioneering accounts, but it is a mark of how far we have come that they can now seem somewhat simplistic and dated. We must remember that these accounts are the ancestors of all the work now done in feminist history of ideas; it is worth looking at the concerns of these accounts in order to trace the genealogy of the types of scholarship being done today. I will take my examples from Okin's pioneering and often very insightful survey.

These feminist accounts of the 1970s and 1980s work on a broad canvas, not afraid to take on a great sweep of Western intellectual history: from Plato to Mill, from Plato to contemporary political thought, or at the very least from Plato to Machiavelli.[19] They are explicitly concerned to criticise the traditional canon of (male) authors, precisely *because* it is a canon and is seen as 'the basis of our political and philosophical heritage'.[20] The relevance of these authors for modern feminism is a prime motivation in studying them: essentially, we are concerned to find out how much of that heritage we need to jettison if we are to achieve equality between the sexes. The authors are read largely for what they say specifically about women, women's nature, and women's role (and how that relates to men's role). Women, rather than 'gender', is the key word; and metaphorical extensions of maleness and femaleness are of relatively little concern. The overall emphasis of the works is on a feminist critique of the canon: Witt has dubbed this stage in the feminist history of philosophy 'negative canon formation'.[21]

In keeping with the broad sweep of material which they cover, these writers are able to come up with at least some overall conclusions about the material they cover. Okin, for example, argues that those thinkers who are committed to a view of the family as 'a natural and necessary institution' end up requiring of women types of behaviour which are not morally prescribed for men. As an extension of this, she finds that many philosophers do not extend their conceptions of 'human nature' to women. Women are treated as functional, rather than as bearers of a full human nature.[22] Not only are generalisations possible over long stretches of intellectual history; we can also draw conclusions for our own times about the nature of a political theory which would meet feminist requirements.

Part of the reason why Okin's accounts are often rather strong is that she does insist on looking at philosophers' views of women in the context of their broader philosophical views. However, this might not be for the same reasons as modern scholars. Okin has two concerns here: to see whether a philosopher's *whole* theory is vitiated by his views of women; and to be able to convict him of inconsistency between his general principles and his views of women. (She focuses, for example, on Rousseau's failure to extend his central political principles of equality and freedom to women; more subtly, she brings out the flexibility of the concept of 'nature' in both Rousseau and Aristotle.)

The obvious critique to be levelled against this early work is the simple one of unsubtlety: in their keenness to right the wrongs of centuries of thought, these scholars naturally focused their gaze very directly on the views of women in the texts, and particularly the most negative aspects of those views. For example, Okin, reading Aristotle, rightly perceives the 'multiple' standard of values he applies to the human beings of different types who exist in the polis, but is more insistent that modern scholars should 'see that the injustice of his treatment of slaves, women and workers is all of a piece' than that we should draw out the implications of this multiplicity.[23] In this example, as in much of the work, the importance of making a judgement about the legitimacy of past ideas seems to get in the way of a subtle explication of them.

So what have the major developments been since these early days of the field? There has been a growth in approaches which do not merely look at the question of 'women' but at gendered discourse more broadly. The concern to convict authors of sexism has given way to a more nuanced attention to complexities in authors' views of women and use of gendered language and symbolism. Changes in the general nature of the subject are no doubt connected with changes in the nature of feminism, as well as with the maturing and consolidation of a new academic field. Feminism within academia – which sadly tends to be less connected with feminism outside it than was the case in the 1970s and 1980s – has many strands, but in general it is less single-minded and revolutionary than was the case when these first accounts were produced. There is a reduced willingness to see male authors as simply perpetrators and beneficiaries of patriarchy; instead, a feminist analysis which sees us all as caught up in the same networks of value, and struggling to articulate ourselves within them in ways which make sense to us, leaves much more room for acknowledging the complexities of both men's and women's discourses. In keeping with that, feminist social historians have emphasised women's agency, and the ways in which women negotiate their lives within the (sometimes limited) options which are available to them. A parallel development in the reading of political texts from the past would lead us to see not monolithic texts against which women could gain no purchase, but the ways in which authors' views of women or of gendered characteristics might be more porous, leaving space within which women can manoeuvre. This is one reason for the striking shift towards more recuperative feminist readings even of the most apparently sexist canonical authors. The feminist purpose of the earlier scholarship has not been lost, but it has moved into reconstruction as well as critique, and there is now a willingness to 'reclaim' or 'appropriate' elements in past thinkers which in some way resonate with feminist concerns. Thus Lloyd's account of the consistent male gendering of reason is followed by Karen Green's insistence that another reading of the tradition uncovers ground for a feminist humanism which does not abandon rationality. Similarly, Lloyd's collection of work in the feminist history of philosophy highlights authors who read texts from the past sympathetically in the light of the modern feminist concerns.[24] Indeed, it has become apparent that the critique of the tradition of political philosophy was partly misguided: the tradition is full of material of great richness which feminists can 'think with' in productive ways. It may rather have been modern traditions of interpretation which led earlier feminists to believe in an austerely male-oriented philosophical and political tradition.[25]

It may be illuminating now to stop and take stock of the current state of scholarship in the area of gender and the history of political thought. It is striking that the sweeping surveys discussed above seem to be a thing of the past. To a certain extent they have their heirs in the more specialised study of particular authors in that canonical history from a gendered perspective; but it is not simply a case of the fragmentation of grand narratives into smaller studies. Rather, there is also an expansion and diversification of the ways in which gendered ideas from the past can be examined. What we have now is a very rich field, but one in which the relationships between different approaches, and the ways in which they could feed into each other, are far from clear.

Much of the work on the canon of political writing has been done by feminist philosophers and political theorists. The movement outlined above from critique of the canon to a more positive construction of feminist political thinking has led to a reduced emphasis in this quarter on interpreting the traditional canon, or at least the earlier parts of it. Feminist work in contemporary political theory and practice has perhaps now got off the ground to such an extent and in such a way that the critique of the thinkers of the past is only relevant in certain limited parts. Such work often pays (relatively) little attention to the canonical tradition, even where it is conscious of itself as arising out of an earlier set of feminist critiques of it.[26] Of course, there is still room for the rehabilitation and appropriation of concepts which are seen as viable for feminist projects, as noted above – although many of these are in the areas of moral theory or epistemology rather than directly political.

Within the history of political thought as a field in itself, gendered work on particular authors and periods has continued to be done. Two of the most influential specialist works, Carole Pateman's *The Sexual Contract*, which offered a strong feminist critique of early modern contract theory and thus of the foundations of liberalism, and Hanna Pitkin's study of Machiavelli, date back to the 1980s.[27] More recent scholars have not produced influential works on a similar scale, and methodologies and standard concerns have also been slow to develop. But material is gradually accumulating, through parts of larger works on particular works or authors, and through articles, now usefully being flagged up and collected in the 'Re-reading the Canon' series *Feminist Interpretations*.[28] The articles in each of these volumes testify to the continued interest in gendered approaches, but also to the (healthy) lack of unity of purpose, as some authors tend towards critique while others offer interpretation of a more neutral kind, or offer readings which are intended as starting-points for new developments in feminist thought. In a sense, too, what has happened is that works focusing directly on gender have been subsumed into work which integrates gender into a broader set of concerns within a text. For example, on Plato, detailed work has been done which integrates themes of gender with consideration of rhetoric, heroic values and manliness, or 'the erotics of statecraft'.[29] With the increasing sophistication in views of the role of gender within texts, it is natural that a focus on 'women' (or even men) should give place to readings of gender which offer a much more subtle understanding of the operation of gender within the logic and themes of the texts.

What about broader trends in the history of political thought? How might they impact on the inclusion of gendered perspectives? Well, recontextualisation and deeper contextualisation, with 'context' being treated as the contexts of inherited and current texts and modes of speech, as well as the political and cultural contexts in which canonical texts were written, should in itself recover the works and thoughts of women who have not been part of the 'canon' but who corresponded with or wrote works in response to men who are.[30] There is also the scope for such contextualisation to draw out the significance of gendered debates which revolved around particular political events. There is, however, little sign of the main stream of historians of political thought taking contextualisation in these gendered directions. The work which does move in such directions tends to pay rather less attention to the canonical authors (which may be no bad thing), and is located at the intersection between the history of ideas and cultural and social history. Two good examples of this kind of work are Lynn Hunt's study of the 'family romance' of the French Revolution, and Rachel Weil's richly detailed *Political Passions*, which studies the gendered discourses of later seventeenth-century English politics.[31]

Most feminist work being done within history tends to focus on social history, often extended into the cultural and intellectual aspects of gender to some extent. In many cases, it is literary scholars who have explored the space left by the relative lack of gendered approaches to political and intellectual history, making a lot of the running in recovering women who wrote in politically significant ways, and reading literary sources (sometimes under a 'New Historicist' banner) in ways which are very sensitive to their political, historical contexts as well as their gender implications.[32] Hero Chalmers's book examining royalist women writers of the later seventeenth century from a literary perspective informed by their royalism is one example of how this may intersect with questions of gender.[33] This kind of work is part of a growing trend, but the concerns of literary scholars can be very different from those of historians of political thought, and how much potential there is for this work to be fed back into such history may be doubted.

One final intersecting area which is of relevance to gender and the history of political thought is the question of how we treat female authors, most of whom have no place in our canon of great thinkers. I have suggested above that, with deeper contextualisation of texts, women writers ought to become more present even in studies of canonical (male) figures. One feels that this process will be slow; perhaps these writers need to be made better-known through separate studies first, and will then be drawn into more mainstream histories. The inclusion of female thinkers in our study of the thought of the past is a matter of justice to which we should be committed as feminists (assuming, as we surely must on the basis of our increasing knowledge of these writers, that the exclusion of women from the canon is not simply a reflection of merit). The extent to which we should expect it to be revealing about the gendered nature of political thought is disputable; we should certainly not force gender identification on texts which do not show them, or select our female canon on the grounds of the female focus of works.[34] The analysis of the beginnings of 'feminist' thought,[35] which is certainly a part of intellectual history in its own right, must also have implications for the history of political thought in the early modern period. Just as male authors' gendered views can be partly constitutive of their political philosophies, so anyone who writes feminist work in a context where political discourse is gendered is almost necessarily making politically significant points.[36]

Having glanced at the range of work being done in the broad field of the gendered history of political thought, we need to pull together some issues of methodology which arise. One standard view of the purpose of the *historical* interpretation of texts in political thought is that it is intended to recover original meanings, which may not be transparent to the modern reader. On this view, the importance of context is that it enables us to understand what the text's author may have been *doing* in saying what she/he did. Through an absorption in the context of the text's production, the historian can provide not a crude biographical motivation which explains the author's writing, but the network of contemporary meanings within which the author's statements can be understood. Quentin Skinner has suggested a test which can be applied to a modern interpretation: could the author be brought to accept that this was the meaning of her/his text?

Such a view of the nature of historical interpretation of texts does not, of course, rule out the significance of gendered perspectives. Rousseau wrote much about the nature and role of women, and just this type of analysis might be helpful to elucidate what he thought he was saying. Having looked at the material about women in his philosophical treatises and fictions, and located it in the context of other arguments about women's role at the time, one could construct an analysis of Rousseau's meaning, travel back in time, and stand a reasonable chance of him agreeing with one's interpretation of his meaning. Again, looking at the more metaphorical presence of gender in these texts, one might interpret them 'with the grain' as ways in which the author was communicating his or her meaning, and come up with an account of their purpose that the author could agree with. However, the study of gender within intellectual history has rarely confined itself to this kind of explication (however demanding this task alone may be).

One reason for this is that the study of gender within the history of ideas has been enabled and largely motivated by feminism. This means that the value-free interpretation of the gendered ideas of the past has to be set against the feminist commitment of the scholarship. Feminist scholarship is likely to be more conscious of its purposes in the present, and the approach to past ideas is likely to involve an element either of critique or approbation. While the crudest forms of value judgement on the gendered language and argument of the past is not a major feature in more recent scholarship, there is likely to be an ongoing commitment to assessing the significance of past ideas (or at least the broad ways in which theories and texts are gendered) for contemporary feminism. Feminist scholars do not pretend that they can abstract themselves from the present and their own beliefs in interpreting the texts of the past. Arguably, this self-consciousness about engagement in the material studied may be more helpful in avoiding inappropriate anachronism and value judgement than a naive belief in neutrality. But feminist scholars should attempt charity in their critique of texts of the past, and be clear about what is being critiqued for present purposes rather than as being inadequate by the standards of the past.[37]

There is a second, separate reason why gendered approaches to the history of ideas tend not to stick to the strict prescriptions of the 'Cambridge School'. This is to do with the ways in which gender is present in texts, which, as we have seen, can be complex and metaphorical. In order to consider the gendered aspects of texts and theories, the historian may have to let go of the hunt for the author's conscious meaning and approach texts and sources in quite different ways. (Whether this involves abandoning a search for the meaning, or meanings, of the text is a question which can be ignored for present purposes.) Structures of gendered thought may certainly be present in ways which do not serve the author's conscious purposes. The interpretation of these structures then has to proceed on a different basis. I will discuss two broad approaches which fall under this heading.

Firstly, there are approaches which attempt to disinter patterns which are present in the sources at a 'deep' rather than a surface level. In this case, it may be very plausible to say that these elements are not part of the intended meaning of the text, though they may not necessarily work *against* that meaning. Examples of this would be interpretations which use psychoanalytical ideas to draw out significant patterns in the sources. Hanna Pitkin, for example, interprets the many powerful and solitary male figures who feature in Machiavelli's writing with reference to psychoanalytic ideas.[38] Similarly, Lynn Hunt's study of the many discourses surrounding the French Revolution in terms of their 'family romance' does draw on many direct discussions about what should be done about the place of women or the structure of family life, but interprets these, and other less direct sources, in terms of broader overarching themes of the overthrow of the 'father's' authority and the consequences for the 'brothers' who now rule. It is unlikely that many of those who generated these sources saw themselves in exactly these terms – Hunt is concerned with the 'collective political unconscious', although explicit familial analogies for politics do form part of her evidence.[39] Pateman is another author who uses psychoanalytical ideas, and again her interpretation of contract theory seems intended to get at something other than the conscious meanings of texts. In order to make her case for the fraternal 'sexual contract' by which the brothers, having overthrown their father, assert their 'sex right' in the women, Pateman has to argue that this important transaction has been silenced in the texts, but that it is the only explanation which can really make sense of them.[40] It is hardly part of the overt meaning of the texts, then, but it is crucial to the interpretation of the texts' significance.

Secondly – and this may overlap – there are approaches which look, not necessarily for 'deep' structures within the text, but at the problematic elements, tensions or instabilities in the text. One of the most serious challenges to the 'author intention' view of the proper historical interpretation of texts arises from deconstructionists' understanding of the unstable nature of language and texts. In deconstructionist readings, the notion of authorial intention is displaced in favour of the 'play' of meanings within a text. Texts do not yield univocal meanings which can be confidently treated as the intention of the text, let alone of the author. Such tensions may often be present around issues of gender – because gender is deeply embedded in the way people think, as we have seen, and anxieties about it may interact with authors' conscious thinking in odd ways. A focus on finding the 'right' historical interpretation of a text might lead to the eliding of such problematic aspects of the text – which themselves may be far more revealing than a cleaned-up de-gendered authorial intention. Penelope Deutscher has argued, using the resources of deconstruction, for a feminist history of philosophy which pays productive attention to the instabilities in texts' treatment of gender. She criticises feminist scholars for naively assuming that an author who states a sexist view but doesn't apply it consistently is not as sexist as all that. She argues that instead we need to develop sophisticated readings of gender instability in texts, and understand that such instability can reinforce the effects of gender. For example, the resilience of notions of gender is upheld rather than undermined by the ability to describe men as feminine and women as masculine. Such flexibility enables these categories to be normatively applied; in Deutscher's terms, such instability is 'constitutive' of gender: 'Gender relies on incoherence of definition.' Applying such insights to a highly ambivalent author such as St Augustine can yield illuminating results.[41]

For all of the reasons discussed, when reading texts for their gendered aspects we are likely to do far more than just look for the authorial meaning of the text. We might want to say that texts have far more meaning(s) in them than simply the authorial intention we might hope to discern. Alternatively, we might want to say that we are as much concerned with the 'effects' of a text as with its meaning: the (varying) effects it produces for all its readers, male and female, then and now. All of these types of gendered reading can be productive, and which is suitable will depend on the nature of the sources as well as on the concerns that are brought to bear on the text. What is necessary is a degree of self-consciousness about what is, and is not, being claimed, and about which aspects of the interpretation are taken to be historical in nature.

I would like to finish by offering an extended example from an ongoing research project of my own. I hope that it will suggest how illuminating it can be to look at political theories through the lens of gender, and how complex the associations can be which lead to a theory's gendered nature. It may also suggest ways in which it is still possible and worthwhile to trace gendered elements in thought over long spans of time, without merely resorting to the persistence of patriarchal tradition as a catch-all explanation. This is because there are recognisable traditions of political thought which are drawn on and developed over centuries; if these are closely correlated with gendered understandings, those can also continue to operate within these traditions. I have approached this project as a historian, seeking to interpret a set of texts historically. I feel that in this case, gender illuminates the nature and intentions of the theory. However, I do also feel that this analysis may reflect back on the issues facing actual women approaching political traditions. In this sense, looking at the gendered elements which are manipulated in the theory is still a feminist project; the abstraction of the idea of gender from real men and women within the theory does not prevent it from having consequences for women and men who encounter it.

The republican tradition in early modern Europe is a major focus of scholarly interest at the moment. It can be hard to locate a single feature common to all republican writings, but essentially they mobilise classical sources, and sometimes the example of existing (Italian) republican states, to argue against the domination of a monarchical ruler and in favour of a participatory form of government. The tradition is strikingly masculine in its language, often making much of its citizens' martial valour in defence of liberty, and almost always insisting on the *virtù* or 'virtue' of its citizens in ways which are given strongly masculine connotations. These writings are not likely to offer a woman reader the sense that she is being addressed, or that engagement in government is possible for her. While it might be possible to put this down to simple sexism and argue that the republican tradition simply valorises itself by emphasising masculine qualities, I think there has to be more to it than that; other political traditions may be equally exclusive of women in practice but not so consistently gendered in their language.

In answering the question of why republicanism is masculine, I have considered the sets of arguments which structured republican thought: those arguments which were available for advocates of participatory government to draw on, but also those which they might have felt vulnerable to and wanted to defend themselves against. This involves looking right back to the classical writers – not only Roman writers like Cicero who might plausibly be considered republican, but also the Greek writers who wrote about democracy. Plato and Aristotle are not exceptional among surviving ancient Greek writers in their hostility to (Athenian or extreme) democracy; we have no direct defences of democracy by ancient authors, and there was a remarkably coherent tradition of critique of democracy.[42] Plato and Aristotle use many of the same argumentative tropes in criticising democracy as other less well-known authors.

My argument is that defenders of republicanism had to show how republicanism could be a good form of participatory government when 'democracy' – in the powerful and influential Greek philosophical tradition, and in early modern usage of the word – was a bad form of participatory government. Were they not really the same thing?

Republican writers tried to distance themselves from the negative characteristics which had been imputed to 'democracy', and one of these characteristics was femininity. Oddly – to anyone who is familiar with the limited nature of the Athenian (native, adult, male, free) citizen body – the anti-democratic writers insisted that democracy gave excessive freedom both to women and to slaves.[43] As well as these claims about the extent of the freedom afforded to women under democracy, Plato makes a more abstract association between democracy and femininity:

> 'It's probably the most attractive of the regimes', I said. 'Like a coat of many colours, with an infinite variety of floral decoration, this regime will catch the eye with its infinite variety of moral decoration. Lots of people are likely to judge this regime to be the most attractive – like women or children looking at prettily painted objects.'[44]

The idea seems to be that – whatever the realities of life under the ancient democracies for women and slaves[45] – for anti-democrats there was something powerfully feminine about these regimes. Even if women and slaves were not in fact citizens of them, they were regimes suited to women and slaves – or to effeminate and servile men. Why should this be?

These anti-democratic authors see democracy as being an incorrect form of government which does not work for the interests of the polis as a whole, or for higher ethical ends, but is simply a mechanism for the fulfilment of the self-interest of the mob. Pseudo-Xenophon makes this entirely clear: 'The people itself I personally forgive its democracy; for everybody must be forgiven for looking to his own interest.'[46] It is the psychological basis of this self-interested action which makes it – and democracy – feminine. When democracy is considered as a psychological trait, as it is for Plato's 'democratic man', this is at its clearest. The democratic man applies democratic political principles – the use of the lot, the insistence on absolute equality rather than meritocracy – to his own desires: 'Putting all his pleasures on an equal footing, he grants power over himself to the pleasure of the moment, as if it were a magistrate chosen by lot.' Rather than selecting morally good desires to act on, he insists 'that all desires are equal, and must be valued equally'.[47] What is striking about this is that the democratic man's soul looks very much like a woman's soul – as Aristotle explicitly depicts it, and as Plato hints it would be. Aristotle makes clear that the souls of women, children and (natural) slaves all lack, in different ways or to different degrees, the effective directive power of reason over the desires.[48] It seems that a democracy is a large-scale model of a feminine or servile soul, where appetites rule over reason.

This set of ideas about democracy crucially affects what proponents of republicanism are able to say in its defence. The texts of Plato and Aristotle, in the original Greek, became increasingly accessible to the Renaissance writers who developed the language and arguments of republicanism, and the anti-democratic, and masculine, tone of republicanism was to some extent already set in the writings of Romans such as Cicero – who was himself very familiar with the Greek writings, and who in his *De Re Publica* chose to insert a translation precisely of Plato's comments in the *Republic* about the licentiousness of democracy.[49] The insistence of these early modern writers on the *virtù* of the republican citizen can be explained as a reaction against the psychological qualities attributed to democracy. The good psychological order of the republican citizen, who exercises self-control and devotion to the common good through the rule of reason over desires, guarantees the distance of republicanism from democracy. Guicciardini, considering the problems of Florence and hoping for a healthier republic, encapsulates the republican view connecting the gendered psychology of desire with political weakness:

> All these bad practices have no useful function at all, not even superficial ones, because they do not fulfil any reasonable needs, but only certain vain and empty urges – they satisfy desires that are to be expected in women rather than in men.[50]

When the republican tradition is taken up by English writers to talk about the situation of the nation after the execution of Charles I, similar themes are again present. Milton, praising in masculine terms the 'fortitude and Heroick vertue' of those who stand for the liberation of the country from tyranny, explicitly draws the parallel between the governance of the soul and the governance of the state which was present in Plato. Those who are tyrannised by their own 'blind affections' 'strive ... to have the public State conformably govern'd to the inward vitious rule, by which they govern themselves'.[51] A complex set of ideas about individual psychology and morality, and their relationship to political life, underpins republicanism's masculine emphasis. Milton's awareness of the specificities of the ancient anti-democratic arguments is clear when he writes – again emphasising the masculine – that a 'free Commonwealth' is 'the noblest, the manliest, the equallest, the justest government, the most agreeable to all due libertie and proportiond equalitie'.[52] Here the masculinity of the republic is lined up precisely with those qualities which distance it from democracy: for the ancient anti-democrats, democracy was at fault precisely in its excessive liberty, and in its belief in equality even for those who are unequal: but Milton's republic displays *due* liberty and *proportioned* equality.

What does this example mean for the interpretation of gendered language in texts? I have suggested that there are powerful polemical reasons why authors wanted to emphasise qualities thought of as masculine, and to represent both republican citizens, and perhaps by extension republican states, as models of masculine self-control and reason. Through such moves, writers could demonstrate that republicanism was as far from democratic licence as true men were from feminine self-indulgence, and could thus defend participatory government against charges often made against it. Does that mean that the use of these arguments is purely instrumental, and that their gendered aspects – important though they may be to understanding the authors' polemical purposes – are merely accidental? I think not. These authors in writing as they do are certainly drawing on some rather persistent ideas about women's psychology to make an argument for a political system which might not, to us, seem to have any necessary connection with masculinity, but by doing so they are also reinforcing and reconstituting ideas of masculinity and femininity in a political and moral context. It's perfectly justified for a feminist critic to be concerned with these ideas; what I hope I have shown is that *any* reader of these texts would do well to pay attention to them.

iii

The study of gender within intellectual history is a field which already boasts much stimulating work, and which has enormous potential for rich and varied contributions in the future. There are some signs that feminist philosophers are beginning to define a particular field through their contributions (although even there, approaches are various), but there are also far more ways in which the area can be approached. We should value this diversity of approach, and as historians we should seek to contribute to it in ways which are attentive to the *historical* dimensions of gender and of intellectual traditions, prioritising the close examination of contexts, both concrete and intellectual, for the sources we study. Such approaches are part of a feminist project of the recovery of the history of gender in its broadest sense, which is a necessary part of writing the history of women and of men. But they are also a way of shedding new light on the complex and sometimes contradictory meanings of texts from the past.

notes

1. G. Lloyd, 'Feminism in History of Philosophy: appropriating the past', in M. Fricker and J. Hornsby, eds, *The Cambridge Companion to Feminism in Philosophy* (Cambridge, 2000); G. Lloyd, 'Introduction' in G. Lloyd, ed., *Feminism and History of Philosophy* (Oxford, 2002); C. Witt, 'Feminist History of Philosophy' in E. Zalta, ed., *The Stanford Encyclopedia of Philosophy* (Winter 2000 edition), <http://plato.stanford.edu/archives/win2000/entries/feminism-femhist/> (including an extensive bibliography of work in the field, compiled by A. Gosselin).
2. T. Laqueur, *Making Sex: body and gender from the Greeks to Freud* (Cambridge, Mass., and London, 1990).
3. D. Riley, 'Does Sex have a History?' in J. W. Scott, ed., *Feminism and History* (Oxford, 1996).
4. For one comprehensive survey from the classical world to the beginnings of the Renaissance, see P. Allen, *The Concept of Woman* (Grand Rapids, Mich., 1997–2002,

2 Vols, 2nd edn). N. Tuana, *The Less Noble Sex: scientific, religious, and philosophical conceptions of woman's nature* (Bloomington and Indianapolis, Ind., 1993) is overly sweeping but covers some of the ground.

5. M. Sommerville, *Sex and Subjection: attitudes to women in early-modern society* (London, 1995).

6. Ibid., pp. 177 (on whether marriage required consummation) and 194 (on separations and remarriages in spite of the near-impossibility of full divorce).

7. See, for example, A. Shepard, *Meanings of Manhood in Early Modern England* (Oxford, 2003), who suggests that the gestures and behaviour of some men can be interpreted as signs of an alternative code of manhood to that found in the prescriptive literature of the time.

8. For a discussion of metaphors of gender, and the way in which they can escape from association with real maleness/ femaleness or even masculinity/ femininity, instead becoming 'symbolic', see G. Lloyd, 'Maleness, Metaphor, and the "Crisis" of Reason' in L. M. Antony and C. Witt, eds, *A Mind of One's Own: feminist essays on reason and objectivity* (Boulder, Colo., and Oxford, 1993).

9. Pythagorean Table cited from S. Lovibond, 'An Ancient Theory of Gender: Plato and the Pythagorean Table' in L. Archer, S. Fischler and M. Wyke, eds, *Women in Ancient Societies* (London, 1994), pp. 88–101. See also G. Lloyd, *The Man of Reason: 'male' and 'female' in Western philosophy* (London, 1984), p. 3.

10. Lovibond, 'An Ancient Theory of Gender'; Lloyd, *The Man of Reason*; C. Witt, 'Form, Normativity and Gender in Aristotle: a feminist perspective' in C. Freeland, ed., *Feminist Interpretations of Aristotle* (Pennsylvania, 1998).

11. G. Lloyd, *The Man of Reason: 'male' and 'female' in Western philosophy* (London, 1984; 2nd edn with new introduction: Routledge, 1993).

12. Here my approach as a historian leads me to disagree with Charlotte Witt's philosophical approach to whether theories are 'intrinsically' or merely 'extrinsically' gendered: Witt, 'Feminist History of Philosophy'.

13. My own interests focus on classical and early modern Western political thought, and these are the fields I will draw on here. No doubt more recent intellectual traditions may raise slightly different questions, having more continuity with current political argumentation and therefore, perhaps, raising more acute questions about feminist responses.

14. Well-known examples of political writing about the legitimacy of female rule are John Knox, *First Blast of the Trumpet against the Monstrous Regiment of Women* (Geneva, 1558); Jean Bodin, *Six Bookes of a Commonweale*, ed. K. McRae (Cambridge, Mass., 1962), pp. 746–7: defence of France's Salic law barring women from rule. Sommerville, *Sex and Subjection*, pp. 51–60 on female rule and English and French views of the Salic law.

15. M. Nyquist, 'Profuse, Proud Cleopatra: "barbarism" and female rule in early modern English republicanism', *Women's Studies* 24 (1994), 85–130.

16. Aristotle, *Politics* 1252b1.

17. Ibid., 1259b1; *Nicomachean Ethics* 1160b22–.

18. M. L. Shanley, 'Marriage Contract and Social Contract in Seventeenth Century English Political Thought', *Western Political Quarterly* 32 (1979), 79–91.

19. S. M. Okin, *Women in Western Political Thought* (London, 1980; orig. Princeton, NJ, 1979); D. H. Coole, *Women in Political Theory: from ancient misogyny to contemporary feminism* (Brighton, 1988); and A. W. Saxonhouse, *Women in the History of Political Thought: ancient Greece to Machiavelli* (New York, 1985), respectively.

20. Okin, *Women in Western Political Thought*, p. 4.

21. Witt, 'Feminist History of Philosophy'.

22. Okin, *Women in Western Political Thought*, p. 9.
23. Ibid., p. 95.
24. K. Green, *The Woman of Reason: feminism, humanism and political thought* (Cambridge, 1995); Lloyd, 'Introduction' to *Feminism and History of Philosophy*, pp. 7–16, and essays by Baier, Homiak and Shapiro; Witt, 'Feminist History of philosophy'.
25. Lloyd, 'Introduction' to *Feminism and History of Philosophy*, p. 9; Lloyd, 'Feminism in History of philosophy' offers a more detailed consideration of how to conceptualise this feminist project of 'thinking with' past philosophers.
26. See, for example, D. Bubeck, 'Feminism in Political Philosophy: women's difference' in Fricker and Hornsby, eds, *The Cambidege Companian to Feminism in Philosophy*, pp. 185–204; N. J. Hirschmann and C. Di Stefano, eds, *Revisioning the Political: feminist reconstructions of traditional concepts in western political theory* (Colorado and Oxford, 1996).
27. C. Pateman, *The Sexual Contract* (Cambridge, 1988); H. F. Pitkin, *Fortune is a Woman: gender and politics in the thought of Niccolò Machiavelli* (Berkeley, Los Angeles, Calif., and London, 1984). I discuss these books further in the discussion of methodology below.
28. Published by the Pennsylvania State University Press, 1994 onwards. Single-volumes cover canonical figures in the Western philosophical tradition.
29. M. S. Kochin, *Gender and Rhetoric in Plato's Political Thought* (Cambridge, 2002); A. Hobbs, *Plato and the Hero: courage, manliness, and the impersonal good* (Cambridge, 2000); W. R. Newell, *Ruling Passion: the erotics of statecraft in Platonic political philosophy* (Maryland and Oxford, 2000). Contrast these with M. Buchan's more pedestrian *Women in Plato's Political Theory* (Basingstoke and London, 1999).
30. For example, Mary Astell and Damaris Masham, discussed in P. Springborg, 'Astell, Masham, and Locke: religion and politics' in H. Smith, ed., *Women Writers and the Early Modern British Political Tradition* (Cambridge, 1998).
31. L. Hunt, *The Family Romance of the French Revolution* (London, 1992) (discussed further below). R. Weil, *Political Passions: gender, the family and political argument in England 1680–1714* (Manchester and New York, 1999).
32. S. Wiseman, '"Adam, the Father of all Flesh": porno-political rhetoric and political theory in and after the English Civil War', in J. Holstun, ed., *Pamphlet Wars: prose in the English revolution* (London, 1992).
33. H. Chalmers, *Royalist Women Writers, 1650–1689* (Oxford, 2004).
34. Selective focus on women's more 'feminine' works is remarked by Joanna Russ, *How to Suppress Women's Writing* (London, 1983), pp. 62–75. Feminist scholars may be colluding in such tendencies by focusing on female authors' more 'feminist' work, as in the cases of Christine de Pizan, Mary Astell and Mary Wollstonecraft: B. Carroll, 'Christine de Pizan and the Origins of Peace Theory', P. Springborg, 'Astell, Masham and Locke' in Smith, *Women Writers*. Witt, 'Feminist History of Philosophy', comments on the lack of any unifying female viewpoint emerging from studies of female philosophers.
35. C. Jordan, *Renaissance Feminism: literary texts and political models* (Ithaca, NY, and London, 1990); H. Smith, *Reason's Disciples: seventeenth-century English feminists* (Urbana, Ill., and London, 1982).
36. The relationships between views of women and political beliefs may, as with non-feminist authors, be very complex. Thus Cavendish and Astell both use the language of consent to query male authority over women, but do not wish to use it to challenge royal power. Both are discussed in Smith, *Reason's Disciples*. Wollstonecraft's views on the position of women and the nature of politics in general

are more consistent – as perhaps implied by her (almost) parallel titles *A Vindication of the Rights of Men* (1790) and *A Vindication of the Rights of Woman* (1792).

37. Margaret Sommerville's charitable assumption of the good faith of the early modern theorists of the natural inequality of the sexes is salutary: Sommerville, *Sex and Subjection*, pp. 5–6.

38. Pitkin, *Fortune is a Woman*, for example chapter 3 on the figure of the 'Founder'.

39. Hunt, *Family Romance*, p. xiii. Hunt uses notions found in Freud's *Totem and Taboo* but looks for them and interprets them in the precise historical context of the French Revolution, finding them in that specific case to be supported by a wide range of evidence (pp. 6–8).

40. Pateman, *The Sexual Contract*, p. 110: 'In the stories of the classic contract theorists the sexual contract is very hard to discern because it is *displaced onto the marriage contract.*'

41. P. Deutscher, *Yielding Gender: feminism, deconstruction and the history of philosophy* (London, 1997), citing pp. 91, 1.

42. J. Ober, *Political Dissent in Democratic Athens: intellectual critics of popular rule* (Princeton, NJ, 1998).

43. Plato, *Republic* 563b: 'As for the relationship of women to men [under democracy], I all but forgot to mention the extent of the legal equality [isonomia] and liberty between them' (all translations from Plato, *The Republic*, ed. G. R. F. Ferrari, trans. T. Griffith (Cambridge, 2000)). Aristotle, *Politics* 1319b27–30: 'the measures taken by tyrants appear all of them to be democratic; such, for instance, as the license permitted to slaves ... and also to women and children' (all translations from Aristotle, *The Politics and The Constitution of Athens*, ed. S. Everson (Cambridge, 1996)).

44. Plato, *Republic* 556c.

45. We should not, of course, assume that citizen/non-citizen status was the only determinant of freedom of action; perhaps there was enough evidence of women and slaves out and about on the streets for the claim about their freedom – in comparison to other Greek city-states – to be comprehensible.

46. Psuedo-Xenophon, *Constitution of the Athenians*, II.20 (ed. and trans. H. Frisch, *The Constitution of the Athenians: a philological-historical analysis* (Copenhagen, 1942)).

47. Plato, *Republic* 561a–c.

48. Aristotle, *Politics* 1260a4–14.

49. Cicero, *De Re Publica* 1.67.2 (ed. J. Zetzel, Cambridge, 1995), translating Plato on the excessive liberty of women and slaves under extreme democracy. The text of the *De Re Publica* itself was not recovered until the nineteenth century, but in the early modern period some quotations from it and a summary of it were known, and Cicero's other works express similar values.

50. Francesco Guicciardini, *How the Popular Government should be Reformed* (1512), in J. Kraye, ed., *Cambridge Translations of Renaissance Philosophical Texts: Vol. 2: Political philosophy* (Cambridge, 1997), p. 231.

51. John Milton, *The Tenure of Kings and Magistrates*, in Milton, *Political Writings*, ed. M. Dzelzainis (Cambridge, 1991), p. 3.

52. John Milton, *Readie and Easie Way* (1st edn), *Complete Prose Works of John Milton*, revised edn, Vol. 7, ed. J. Ayers (New Haven, Conn., and London, 1980), p. 359.

11

the politics of intellectual history in twentieth-century europe

duncan kelly

Reviewing so much more than Professor Wegele's *Deutsche Historiographie* in the first edition of the *English Historical Review* of 1886, in what has since become a justly famous essay, Lord Acton critically evaluated various German 'schools' of history in the nineteenth century. Of course, although Acton was the most cosmopolitan of scholars, the fact that the first edition of what would quickly become the premier English historical periodical should be so concerned with the state of German scholarship suggests something more than passing historiographical interest. For at root, Acton's discussions of these schools of history were underpinned by an account of the political implications of different versions of historical enquiry. Indeed, he even suggested – in what had by 1886 already become a standard trope of political discourse – that although German 'historical writing was old', strictly 'historical thinking was new in Germany when it sprang from the shock of the French Revolution'.[1] In conclusion, as well as fulfilling his aim to outline the ways in which these schools 'break new ground and add to the notion and the work of history', the general tenor of the essay offered nothing less than a full-blown evocation of the historical spirit of the age:

> The tendency of the nineteenth century German to subject all things to the government of intelligible law, and to prefer the simplicity of resistless cause to the confused conflict of free wills, the tendency which Savigny defined and the comparative linguists encouraged, was completed in his own way by Hegel.[2]

Given his predisposition towards moral certitude when analysing historical problems, it is quite unsurprising that Acton further argued that 'the marrow of civilised history is ethical not metaphysical, and the deep underlying cause of action passes through the shape of right and wrong'.[3] And although he disagreed with the methodologies of the various German historical schools, whether inductive or deductive in operation (particularly those represented by Karl Friedrich von Savigny and Reinhold Niebuhr, though less so Ranke), a concern with the 'underlying causes of action' nevertheless remained. This kind of moralising historical vision was not new, and in fact exemplified a variation on what had become a central theme of German historical self-understanding by the middle of the nineteenth century. This was, of course, the particular triumph of 'historicism', and the explicit meaning of such a self-consciously historical turn of mind was perhaps best captured in Ernst Troeltsch's famous discussion of just a few years later. Troeltsch suggested that historicism entailed 'a fundamental historicization of all our thinking about mankind, its culture, and values'.[4] However, whilst seeming to present a rather tautological relationship between a concept and its object, this understanding of historicism in fact built on long extant traditions of German political thinking. These were traditions, as George Iggers has consistently reminded his English speaking audience, which stretched back well before the middle of the century and could be found clearly in the historical, political and social writings of Herder, for example.[5] Equally, the actual concept of historicism, or more strictly historism, was not the relatively recent vintage of 1879, as the intrepid Friedrich Meinecke suggested in his famous book on the origins (much older, as he rightly noted) of historicism as a mood. Rather, the concept itself begins much nearer to the beginning of the nineteenth century, perhaps with Novalis's notes from his student days in Heidelberg in 1798.[6] Historicism was an approach designed to stress the importance of historical understanding in the development of individuality and uniqueness; and this, of course, placed it within the broad sphere of *Bildung*, or cultivation. Such cultivation could refer to both the individual self within a particular community, or to the nation. Moreover, whilst the traditionally 'Romantic' merging of the twin characteristics of individual creativity and interdependence associated with *Bildung* was clear, what is often forgotten is how far the Romantic use of this concept had crucial theological precursors and underpinnings, whose ramifications lived on long after the disintegration of Romanticism itself. Indeed, although Madame de Staël's *De l'Allemagne*, which brought the term 'Romanticism' to German educated discourse, neglected this facet, she clearly understood the early German Romantic writers to be engaged in an historical process that something like historicism could both interpret and explain.[7]

Historicism, then, was much more than just the development of a new historical methodology. Unsurprisingly, though, professions of historicism as the most appropriate form of historical enquiry were not at all rare, and claims to write properly 'historical' texts were apparent in most of the major programmatic statements of leading members of particular historical schools.[8] It was seen, quite naturally, as a development of earlier French movements towards source criticism and critical editions that burgeoned in the sixteenth and seventeenth centuries during the broad transition from an 'empirical' to a 'critical' history. Such thinking could then be combined with the development of German ideas about the professorial seminar, *Quellenkritik*, and the continuation of Leibniz's fundamental work as a collator and editor as well.[9] Yet, historicism not only permitted an ethical response to the results of historical investigation, it practically presupposed *a priori* ethical commitments (often, it seemed, geared towards the 'value' of the nation-state, however so conceived), on the part of the investigator. This is certainly one reason why historicism can be so clearly linked to politics, why the rise of history in the nineteenth century is so often explained with reference to the rise of nationalism, and why at least in part numerous different discourses could be united under its banner even though historicism itself was far from being a unified movement. Indeed, *wie es eigentlich gewesen*, Leopold von Ranke suggested that although the study of history could treat of origins and causes when properly executed, remedies for the moral improprieties so worrying to both him and Acton would have to be sought in the new 'science of politics'. Such a science was a relatively recent invention in Germany, thought Ranke, at least at the time of his Inaugural Lecture in 1836, although in so presenting the case he neglected a huge swathe of German political thought concerned with the science of government and human happiness, or *Cameralwissenschaft*.[10] In the lecture, he suggested that politics and history should be separated in practice, whilst the metaphysical underpinnings of his belief in an ordered historical process allowed him to develop a political theory based on universal harmony and the guarantee of an eternal return to order and equilibrium.[11] So, although the objective historian could separate politics and history by the use of a properly 'historical' method, this separation was only ever really a façade because historicism itself had in built presuppositions that were capable of being used to justify multifarious political positions. The development of historicism and its political uses in twentieth-century Europe shows just how diverse a range of positions could be considered under this label, and further illuminates the wide variety of approaches to the study of the history of ideas more generally.[12]

i

At the beginning of the twentieth century, therefore, questions of historicism, the relationship between history and politics, and the political uses of history, were closely intertwined and capable of numerous interpretations. In Germany, Max Weber published what has become a seminal essay on the notion of objectivity (*Objektivität*) in research in the cultural sciences. His ultimate argument suggested that claims to objectivity as some sort of 'neutrality' were chimerical. What objectivity meant to Weber was the recognition that the values of the researcher, whatever type of work was being undertaken, would always inform the research by guiding the particular questions that would be asked, and, ultimately, the way that they would be answered. The most desirable manner of dealing with this, then, was to make one's aims and values in terms of the research as clear as was possible in defining terms, specifying questions and aims, and offering reasoned arguments that might use 'ideal type' constructs, but which were nevertheless fully historically grounded.[13] Of course, the idea that Weber's arguments really require a fastidious extirpation of values from scientific work has been a touchstone of much sociological interpretation of Weber for the majority of the twentieth century, even though such an account could equally well be seen as an intellectual and interpretative red herring.

His celebrated distinction, between an ethics of disposition (*Gesinnungsethik*) and one of responsibility (*Verantwortungsethik*), guided his own conceptions of what was necessary to balance one's life according to the requirements of particular spheres of life in order to become a creature of 'personality'. Intellectual integrity required one to recognise the connections and contradictions between these twin poles of attraction, and not to think that adherence to one alone would be satisfactory; unless of course one countenanced a complete and thoroughgoing religious rejection of the ethics of this world in favour of another.[14] Such claims would have particular importance for Weber's relationship with a scholar often perceived as a sort of alter ego, the doctrinaire socialist and syndicalist Robert Michels. Weber supported Michels's academic career, though tried to push him to recognise how his own values were guiding what he saw as his objective works of social science, particularly his early analyses of socialist parties in Europe which Weber published in the *Archiv für Sozialwissenschaft und Sozialpolitik*, and also of course in his still widely read work, *Political Parties*. But although this intellectual probing forced Michels to clarify his own position at various stages, it is also important to note the impact on his writings of his personal and intellectual engagement with the works of Gaetano Mosca and Vilfredo Pareto, writers typically associated with the development of an 'elitist' theory of politics which suggests that political rule is always simply the circulation of particular groups of elites.[15] Such a theory, presented as 'objective' social science based on empirical observations from history and of human nature increasingly underpinned Michels's move towards his understanding of the behaviour of political parties as based on an 'iron law of oligarchy'.[16] For the development of intellectual history, however, the elitist theory of politics would have wider ramifications, especially in the Anglophone world.

One of the principal reasons behind the development of a particularly historical approach to the study of political theory in post-war intellectual history was a reaction against the idea that historical study was useful to social science and the arts and humanities only to the extent that universal or law-like generalisations could be inferred from its premises.[17] This legacy of elitist theories of politics quickly merged with high economic theories of democracy and social choice on the one hand, and behaviourist social science on the other, leading not only philosophers and historians like Isaiah Berlin and Peter Laslett to wonder whether political theory was dead, but it also prompted questions from mainstream political scientists like David Easton.[18] It was, moreover, important to the rise and decline of what Peter Novick has termed the relatively short-lived 'noble dream' of objectivity understood as value-free investigation in North American historiography, a dream that was quickly unpicked by developments in feminist and Marxist scholarship. This soon resulted in the situation of there being 'no King in Israel', where everyone did 'what was right in his own eyes'.[19]

But the wider interest of this, in Richard Tuck's assessment of the rise of the historical approach to political theory in the post-war period in Anglophone scholarship in particular, is the way in which intellectual history had to define itself firstly against the types of interpretative limits outlined by these arguments, and which he in fact sees as a rather degraded form of Kantianism focused on the distinction between facts and values although without Kant's transcendental ethical theory. Equally, however, Tuck suggests that it has come to be as much defined against the considerably more sophisticated Kantian-inspired theory of later decades instantiated by the work of John Rawls in particular, and instead aims to focus on the necessary relationship between language and political action, thereby suggesting that the history of political thought should be understood historically in precisely the same way as any other component of the human sciences.[20] This had already been argued several decades earlier by Collingwood, who had suggested that all history 'was the history of thought', but it was reprised in this new context for a new generation of intellectual historians and in a new context.[21] For this was a context in which the study of intellectual history was often derided as 'cant', 'flapdoodle', or as the study of lazy ideological *post facto* rationalisations of actions that were always already determined by high-political motives and concerns in the first place.[22] The so-called 'return' of grand theory, whether in narrative historical form or understood as the intertwining of social and political theory, built upon similar foundations and frustrations with the idea that the humanities would be best modelled along the lines of the natural sciences.[23] Rebutting these criticisms and developing the study of intellectual history was then an important motivating factor behind post-war developments in the study of the history of political thought and intellectual history.

However, the relationship between Kantianism and intellectual history took a somewhat different form in early twentieth-century Germany. If Weber's was one attempt – often itself somewhat problematically described as neo-Kantian – to illustrate the foundations of historical research, another lay with the development of forms of legal positivism. On the other hand, the historicist appropriation of Kant for the field of the human sciences, or *Geisteswissenschaften*, developed most self-consciously in the writings of Wilhelm Dilthey. The question of how philosophers returned 'back' to Kant's legacy in nineteenth- and twentieth-century European intellectual history is a vast terrain, and lies somewhat outside the scope of this discussion.[24] But Dilthey's vision of the empathic understanding of other cultures, a philosophy that he developed in part from Kant's thought, informed both his epistemology and his understanding of the nature of modernity. It provided a matrix of differences between particular worldviews that could be studied by hermeneutics. And for Dilthey, this was an historical problem; for later exponents of hermeneutics like Gadamer, it would become an ontological one that required a clear recognition of the interrelationship between historical understanding and contemporary life.[25] Moreover, whereas Kant's ethics crucially returned him to the Greeks, Dilthey's account of modernity was part of an early twentieth-century revival of the study of neo-Stoicism in European intellectual history that would have profound intellectual and indeed political consequences.[26] Dilthey examined in particular Justus Lipsius's combination of political motivations and academic philological scholarship as a key moment in the transition from a Europe mired in religious wars to a more recognisable Europe of nation-states and the rise of modern liberalism. He thus opened up two fields simultaneously: firstly, the intellectual history of early modern Europe, and secondly, an almost teleological vision of the rise of modern liberty with Hugo Grotius as a central figure in the transition from Stoic reason to Protestant natural law.[27] Thus, whereas Weber simultaneously tracked the rise of a particular character type and its elective affinity to the rise of occidental capitalism in his essays on the Protestant 'ethic', Dilthey turned instead to the broadly Protestant tradition of natural law to illustrate an equally broad transition to 'modernity'. The argument still has considerable resonance for contemporary intellectual histories of modernity,[28] though as Peter Miller notes, more recent histories of neo-Stoicism often neglect its deep-rooted connection with religion, an area that earlier twentieth-century scholarship had in fact emphasised.[29]

ii

As Miller goes on to discuss, furthermore, in the prelude to and aftermath of the Nazi period, the writings of Gerhard Oestreich and Otto Brunner in particular in Germany attempted to transform the religious and therapeutic elements of neo-Stoicism as a philosophy of life and as a political stance, into a neo-Stoicism understood as a philosophy of *raison d'état* concerned with the rise of a new and disciplinary society. In a bravura piece of intellectual genealogy, Miller emphasises just how intertwined was the relationship between Brunner's and Oestreich's historical work and their post-war writings that were filtered through an awareness of their position during the period of the *NS-Zeit*. Many other studies have noted the important and disturbing connections between

National Socialism and academic scholarship.[30] But he equally notes that the danger of such positions was recognised by contemporary European scholars of equal weight and stature, such as Arnaldo Momigliano and Fernand Braudel, who respectively challenged both the conservative implications of Brunner's thesis and the one-sided reading of Lipsian stoicism as part of the move towards a disciplinary society by Oestreich. Momigliano in particular noted that such post-war scholarship – with its movement towards a structural or sociological history – was an attempt to efface earlier ideological stances.[31] But the idea of a structural and yet comparative history of society and ideas, nevertheless, came to be associated with writers like Marc Bloch and Fernand Braudel, whilst much subsequent work in this area developed in various ways related to these fundamental thinkers in France, and is typically referred to as the 'Annales' School or approach. Bloch's comparative history of European societies, outlined methodologically in a seminal paper and given full form in the remarkable early study of the so-called 'royal touch', illuminated brilliantly the radicalism of the new methods of historical scholarship that he, along with Lucien Febvre, had developed in Strasbourg earlier in the twentieth century.[32]

This was work that would be modified and developed by Braudel in particular – with a particular broadening of the importance of geographical analysis that was begun by Febvre – in his seminal account of the Mediterranean and in various essays on historical method.[33] This would continue to influence French academic history until the revolts of 1968 (in fact, Febvre's early writings on geography even appear to have impacted upon Ernest Barker's discussion of national character in England in the second decade of the twentieth century).[34] After this time, a putative third phase of 'Annales'-style scholarship came to predominate, though which would develop long-standing concerns about the study of history as the study of *mentalités* and cultures.[35] Its development points towards various fields now much more familiar as 'cultural' history. And in the French case, this has led to fundamental reappraisals of the French Revolution and French intellectual history from explicitly non-Marxist scholarship in the case of writers like François Furet, and towards cultural histories of scholarship, reading and the book, popular history, festivals and memory, in the writings of such thinkers as Robert Darnton, Roger Chartier and Pierre Nora, among others.[36]

Much of this writing, in French and in translation, has transformed Anglo-American scholarship on modern French history.[37] And in an essay on the politics of intellectual history in twentieth-century Europe, the anti-Marxist polemical intent of these writings perhaps bears further emphasis.[38] This seems certainly appropriate in the institutional and political conflicts over the interpretation and legacy of the French Revolution, for so many years during the twentieth century dominated by the grand Marxist narratives of writers like Georges Lefebvre and Albert Soboul, which has moved instead towards a noticeably linguistic and cultural approach to the study of the

Revolution, and to Tocqueville as an interpreter of it.[39] Furet then made this assessment of the Revolution into a prolegomenon, almost, for a more general critique about the political failures of Marxism, or rather communism, in the twentieth century that had represented another grand illusion of particular intellectuals, including Furet himself as a younger man.[40] There is not a little irony, therefore, in the fact that intellectual history itself is seen by many still as a rather marginal sub-field of history and historiography, despite the claims of revisionist intellectual historians, precisely because of the huge impact of Marxist and materialist explanations of the relationship between ideas and social action in the twentieth century. Indeed, whilst political historians like Lewis Namier and Geoffrey Elton derided intellectual history for its apparent ability to lose touch with the real world, they were in turn equally sharply attacked by Marxist writers who wished to assert the strong interdependence between material circumstances and the development of ideas. The group of Marxist historians in the UK that arose out of the context of the New Left were a prime illustration of this.[41] And this tense relationship between intellectual history and Marxist history has helped to define the boundaries of much post-war thinking about how to write the history of ideas in one way or another.

However, attempts to write a social history of ideas focusing on the political and social position of classic authors, typically though not always associated with Marxist theorists, were actually subject to numerous differently focused attacks in the post-war period.[42] For example, scholars such as Darnton have emphasised the rather patrician style of intellectual history that the social history of ideas can lead to. Using Peter Gay's celebrated series of histories of the Enlightenment (which could hardly be read as Marxist), Darnton complained about the ways in which great books and canonical authors were routinely portrayed in the absence of a real sense of intellectual context. In the general absence of any appropriate 'grubbing' around in the archives, their true social situation could not be explained; nor does such an approach show an awareness of the history of reading or books, something that could legitimately be analysed quantitatively in terms of intellectual history.[43] Similar criticisms might well be made about the equally polemical and anti-Marxist intent that continued to help structure the type of history of ideas that one found in the sparkling writings of Isaiah Berlin.[44]

But these are slightly different sorts of critique than those offered by the attacks on historical determinism and class-based history suggested by Furet among others, and different still to the anti-Marxist critique of the social history of ideas offered in most post-war Anglo-American intellectual history and history of political thought that began to be outlined by writers like Quentin Skinner, John Dunn and John Pocock, and which is discussed in other chapters of this book.[45]

Methodologically, this critique has betrayed distaste for the idea that either 'ideas' *sui generis*, or the particular ideas of certain writers, could be read off against a matrix of unchanging questions and perennial problems, practically as if all the 'classics' of political theory really were engagements with a set of timeless questions that could be understood purely textually as many earlier histories of political thought had suggested.[46] It has equally betrayed a withering dissatisfaction with mainstream accounts of certain key authors generally regarded as having canonical status in the history of political thought, and in British political thought in particular. The Marxist, or *marxisant* reading of Locke and Hobbes as 'possessive individualists' and tacit supporters of a nascent bourgeois capitalism in England, the understanding of which provided the key with which to unlock and interpret their writings, was one important illustration of this in a series of polemical critiques.[47] In fact, the political uses of argument during the seventeenth-century English Revolution have, in the study of intellectual history and the history of political thought in the post-war period, long been dominated by clashes between the type of social history of ideas and the avowedly historical study of ideas in context. Nevertheless, the celebrated and cerebral debunkings of the mythical character of such analyses of particular figures and tropes of intellectual history have now, as Thomas Kuhn argued about revolutionary new paradigms of research, become mainstream. In turn, they have also often been systematically distorted.[48]

iii

Yet the idea of writing the history of political thought historically, and having an appropriate historical method with which one can interpret texts, seems in reality to have been the rather modest aim of what ultimately amounted to a root and branch re-evaluation of the study of the history of political thought in the Anglophone world. Only in the Italian tradition, it seems, were there dominant exponents of something approaching the social and cultural history of ideas that fused context with an intellectual attention to particular details without bringing forth anti-Marxist critiques, and that was again to be found in the path-breaking work of Arnaldo Momigliano on the legacy of classical scholarship in early modern and modern historiography specifically, and in the writings of his fellow countryman Franco Venturi.[49] A different illustration of the importance of the social history of ideas to the development of the history of political thought can be found in the writings of Carlo Ginzburg, for example, whose method of 'microhistory' brings together an attempt to illustrate individual and discrete moments (especially in the lives of particular individuals) to promote a sense of shared and empathic understanding. In his own retelling of the history of microhistory, Ginzburg notes that this really began to flourish as a mainstream historiographical paradigm under the imprimatur of Braudel as the 'culmination of the functional-structural approach'.[50] It would also help to inform post-war developments in German social history and the development of 'everyday history', or *Alltagsgeschichte*.[51]

Other authors have developed this line of argument, suggesting that microhistory is, rather, a part of a broader trend towards a socio-cultural history which encompasses intellectual history; a claim denied by those who wish to maintain a sense of specificity concerning what the 'intellectual' in 'intellectual history' actually constitutes.[52] Ginzburg has drawn attention to questions of interpretation and art history to help illustrate the importance of thinking about methods in intellectual history, and in particular to draw out the focus of his own ideas about the 'transmission' of ideas between and within periods and people. And he has illustrated this recently in a full iconographic prehistory of Lord Kitchener's image on the recruitment poster 'Your Country Needs You',[53] and such work shows the intellectual history of politics in various guises as much as it illustrates the politics of intellectual history itself. Moreover, a focus on the longevity and multiple legacies of the classical heritage importantly informed a series of researches and interests that were fundamental to the general history of intellectual history in twentieth-century Europe, and this was a result of the impact of émigré scholarship in general.

The forced migration of many scholars from Nazi Germany has been told in various places and is often concerned with those writers associated with the Frankfurt School, though the overall story of their cumulative impact on the development of twentieth-century intellectual history still waits to be fully recounted satisfactorily. But the development of critical and historical approaches to the study of Fascism pioneered by such émigré scholars built extensively on the early structural analysis of the Nazi dictatorship offered by Franz Neumann in particular. Indeed, whilst one of the emphases of this chapter has been the tense interrelationship between intellectual history and Marxism, the politics of intellectual history in twentieth-century Europe could equally be seen as a structural reaction against political authoritarianism more generally. Neumann himself had tense relations with the members of the Frankfurt School, but developed an analysis of the polycratic structure of what he saw as the 'non-state' of Nazi Germany that combined an analysis of the empirical situation of contemporary Germany with pertinent debates about the rationality of the Western tradition of political and legal thought that took place during the Weimar Republic. This came through especially clearly in his profound engagements with the works of Max Weber and Carl Schmitt as much as with his contemporary social democratic heritage.[54] The grand synthesis outlined by Neumann cannot be understood without an appreciation of these earlier contexts, and nor can the attempt of critical theory to offer a totalising critique of contemporary society be fully explained. Indeed, the work he began under Weimar transferred with him to England, where he studied with Harold Laski and Karl Mannheim, and formed the background to his account of National Socialism in *Behemoth* as well as becoming foundational for later studies more obviously related to the history of modern political thought and the rule of law.[55] Such studies were to have considerable impact on the development of political science in North America by developing analyses of writers deemed important to the development of liberal democracy.[56] Other authors engaged with Weber and Schmitt instead turned towards the rational explication of what were seen as the esoteric writings of classical scholars, an approach associated with Leo Strauss most obviously.[57]

And whilst social theorists have continued to build on the idea of a negative dialectic at the heart of an Enlightenment project in their analyses of the nature of modern selfhood, technology and society, the work of Jürgen Habermas in particular has turned away from the historical importance of the writings of Weber and Schmitt to the origins of these debates. Such elements of intellectual history appear to have been downgraded in his major project of combining normative social psychology and political theory to lend support to his theorising about constitutional patriotism.[58] Schmitt's legacy, nevertheless, has continued to vex Habermas, while the deep-rooted attraction by many of post-war Germany's most illustrious thinkers to his work and his person has not gone unnoticed.[59] Although it is usually as a critic of liberalism that Schmitt has been understood throughout the majority of the post-war period, his impact on the writings of Otto Brunner, Werner Conze and Reinhart Koselleck in particular should not be discounted, especially in the context of this discussion of the politics of intellectual history, given the profound impact that the development of a conceptual history or *Begriffsgeschichtlich* approach has had upon intellectual and socio-cultural history.[60]

Similarly, the writings of Michel Foucault on the appropriateness of an historical and 'archaeological' or 'genealogical' approach to the study of what he saw as particular regimes of truth certainly also utilised the legacy of critical theory, but challenged it in various ways.[61] Yet Foucault too built upon ideas propounded by Oestreich about the origins of a disciplinary society, which ties him back to a concern with the place of neo-Stoicism in the development of modernity, whilst his researches on sexuality, madness and civilisation, the prison, as well as freedom of speech and, latterly, political liberty as a technology of self-control and practice, illustrate a structural focus on practices of 'governmentality'.[62] This included a critical questioning of the nature of the Enlightenment itself, but as Habermas at least implicitly conceded in his polemical confrontations with Foucault over notions of reason and subject-centred activity, as well as those confrontations over the legacy of the Enlightenment and modernity, the value of the genealogical method of investigation was one that could reap much fruit, even if he was effectively trying to dethrone its predominant practitioner. An equally shared concern with practical reason in political deliberation illustrated a common set of interests at a profound and foundational level between the two.[63] But Habermas's concerns with Foucault's essays, given most profound form in the lectures that make up *The Philosophical Discourse of Modernity*,[64] were quickly caught up in practical historical disputes soon after their publication in Germany, and can be read against the backdrop of wider political controversies in post-war German intellectual life – particulary Habermas's role in what became the so-called 'historians' dispute', or *Historikerstreit*.

iv

This dispute, of course, concerned among other things the ability or otherwise of conservative historians to adequately 'come to terms with' the legacy – indeed, in particular the 'image' – of the German past.[65] This was central to his angry clashes with conservative historians like Andreas Hillgruber and Ernst Nolte, and has remained foundational for the development of his ideas about a post-national attachment to liberal constitutional ideals predicated upon theories of post-conventional identity.[66] In terms of the politics of intellectual history, however, there were perhaps two principal implications of these various post-war developments in French and German academic circles that again relate to an attempt to overcome authoritarianism. The first was a calling into question of traditional ways of conceiving of history itself, focusing on the moral implications of historicist approaches to political events, on the one hand, and the practice of writing history itself, on the other. The second was an increased focus on language and discourse in the writing of history, from the aesthetic and moral character of narrative structure in historical writing, and which encompassed the revival of narrative itself and the more general 'linguistic turn' that had such profound effects on the writing of history *tout court* in twentieth-century Europe and which is discussed elsewhere in this book. All of this has presented various possible routes for intellectual history to take.[67] Hermeneutics and its relationship to the recovery of the 'intentions' of texts and their authors have, of course, been a key theme in much discussion of these methods of intellectual history.[68] But such questions have also had particular import for discussions in philosophy over the nature of truth and truthfulness. Bernard Williams's last statement on this issue for example presented a powerful attack on the idea that truth is a concept that can be dispensed with, arguing that if we are to understand the notion of truth philosophically we need also to engage in intellectual history as a form of Nietzschean genealogy to explain it.[69] If intellectual history is to be historical, it must also aim to be truthful in this sense, and this is a claim whose political implications are quite profoundly important for understanding some of the twists and turns of twentieth-century intellectual history.

The moral implications of historicism equally well informed a further strand of scholarship that influenced the politics of intellectual history in twentieth-century Europe, and that once more concerned developments in classical scholarship. The politics of Aby Warburg's distinctive methodological assumptions in his work and in particular the late *Mnemosyne* project – to find a 'basic vocabulary, the *Urworte* of human passion' – were transplanted to London in the form of his original Hamburg library and saved from destruction under National Socialism. This was a physical expression of his promotion of a 'psychological' approach to the explanation of traditions building on the famous library organisation with its 'law of the unknown neighbour' as a guide to building scholarly associations and affinities.[70] The study of iconology and the production of the *Journal of the Warburg and Courtauld Institutes* devoted to the legacy of the classical tradition have, in their turn, promoted key discoveries in the field of intellectual history and the history of art, literature and political theory especially.[71] The journal also provided an outlet for the work of many gifted émigré scholars, again including Momigliano, but also notably Nicolai Rubinstein[72] and the early writings of Ernst Kantorowicz,[73] who offered path-breaking works on Florentine political thought and medieval questions of justice and kingship. Whilst the field of art history was further transformed with the work of Warburg's protégés and students too numerous to mention, Fritz Saxl and Erwin Panofsky in particular began to develop Warburg's studies in a focus on classical mythologies in medieval art, which soon expanded considerably with both scholars achieving considerable renown, and later polymath explorations by Edgar Wind and Ernst Gombrich continued Warburg's legacy.[74]

And while all of this might seem somewhat tangential to a series of reflections on the politics of intellectual history in twentieth-century Europe, the conditions under which intellectual history developed and its focus upon different disciplinary fields nevertheless all coalesced around questions of the historical meaning and interpretation of distinctively intellectual phenomena, whether it be texts, or images, or indeed both.[75] And this has been intimately bound up with political considerations from the very beginnings of intellectual history, and indeed from the very beginning of the writing of history itself. Whether this has been influenced through traditional and obvious forms of loyalty such patronage, whether it has had to do with explicit or implicit political loyalties, or whether it has simply been a recognition that questions of intellectual history never really do have a solely antiquarian interest, that intellectual history is itself intensely political has never really been in doubt. This is so even if – perhaps especially if – the subject matter of many studies appears to be particularly abstruse or arcane. Because at their best such studies always offer us a way of seeing our present through the lens of particular opportunities or moments that were either foundational to the development of a particular field, or they highlight paths not taken, for reasons both good and ill. This is a function of the political nature of the investigation in the first place, though it is not to claim, contrary to some of the critical remarks made earlier, that a simple social history of ideas can hope to make sense of why certain people at certain times choose to investigate particular periods, or issues. Indeed, one of the interesting questions raised by the study of intellectual history is precisely how the historical study of particular works can be said to have some kind of direct import or corollary into discussions of contemporary moment, and how that relationship can be understood in both theory and in practice. For it is not clear that even the most sophisticated discussions of methods have even yet still come to terms with this fact in the actual writing of intellectual history itself. But what this does mean, as Weber, Collingwood and Foucault in modern times well knew, but which has been a commonplace in Western historiography at least since Thucydides, is that the interrelationship between interests, values and narrative structure is itself intensely political. Intellectual history in twentieth-century Europe, a century rightly termed the 'age of extremes', remained intimately bound up with this fact, and there is no reason to suppose that the twenty-first century will be any different.

notes

1. J. E. E. D. Acton, 'German Schools of History', *English Historical Review* 1 (1886), 7–42 (8). The longevity of the analysis can be seen even in the closing pages of recent works such as Harold Mah, *Enlightenment Phantasies: cultural identity in France and Germany 1750–1914* (Ithaca, NY, 2003), pp. 178ff.
2. Acton, 'German Schools', 42, 19.
3. Ibid., 20.
4. Ernst Troeltsch, *Der Historismus und seine Probleme* (Tübingen, 1922), *Gesammelte Schriften* Vol. III, p. 102.
5. George G. Iggers, *The German Conception of History: the national tradition of historical thought from Herder to the present*, revised edn (Hanover, NE, 1988 [orig. 1968]). See also F. M. Barnard, *Herder on Social and Political Culture* (Cambridge, 1969), pp. 39–43.
6. Cf. Friedrich Meinecke, *Die Enstehung der Historismus* (*Werke*, Vol. III) ed. C. Hinrichs (Munich, 1959), trans. J. E. Anderson (rev. H. D. Schmidt) as *Historism: the rise of a new historical outlook* (London, 1972). For further details, see Erich Rothacker, 'Das Wort Historismus', *Zeitschrift für deutsche Wortforschung* 1 (N. F.) (1960), 3–6, especially 4.
7. See John Clairborne Isbell, *The Birth of European Romanticism: truth and propaganda in Staël's 'De l'Allemagne', 1810–1813* (Cambridge, 1994), especially pp. 124ff; for wider reflections on the interrelationship between Romanticism and historicism in Germany, and Berlin in particular, see John Edward Toew's recent study, *Becoming Historical: cultural reformation and public memory in early nineteenth-century Berlin* (Cambridge, 2004).
8. See Keith Tribe, *Strategies of Economic Order* (Cambridge, 1995), pp. 66–94.
9. For the French origins of a 'critical', rather than an 'empirical' history, see, most recently, Jacob Soll, 'Empirical History and the Transformation of Political Criticism in France from Bodin to Bayle', *Journal of the History of Ideas* 64 (2003), 297–316; *Publishing the Prince* (Michigan, 2005); Anthony Grafton, *The Footnote: A History* (London, 2003), especially pp. 122–47; Donald R. Kelley, *Foundations of Modern Historical Scholarship: language, law, and history in the French Renaissance* (New York, 1970).
10. Leopold von Ranke, 'Über die Verwandtschaft und den Unterschied der Historie und der Politik: Eine Rede zum Antritt der ordentlichen Professur an der Universität zu Berlin' [1836] reprinted in Leopold von Ranke, *Sämtliche Werke* (Leipzig, 1867–90), Vol. iii, pp. 280–93. See also the helpful discussion by Gangolf Hübinger, 'Historicism and the "Noble Science of Politics" in Nineteenth-Century Germany', in B. Stuchtey and P. Wende, eds, *British and German Historiography, 1750–1950:*

traditions, perceptions and transfers (Oxford, 2000), p. 192. On *Cameralwissenschaften*, compare the classic treatment of Hans Maier, *Die ältere deutsche Staats- und Verwaltungslehre* (2nd edn, Munich, 1980) and the more recent assessment by David F. Lindenfeld, *The Practical Imagination. The German sciences of state in the nineteenth century* (Chicago, Ill., 1997).

11. See Iggers, *German Conception of History*, especially pp. 78–89.

12. See Richard Ashcraft, 'German Historicism and the History of Political Theory', *History of Political Thought* 8 (1987), 289–324.

13. Max Weber, 'Die "Objektivität" sozialwissenschaftlicher und sozialpolitischer Erkenntnis' (1904), *Gesammelte Aufsätze zur Wissenschaftslehre*, ed., J. Winckelmann (4th edn, Tübingen, 1973), especially pp. 157, 154, 182–5, translated as '"Objectivity" in Social Science and Social Policy', *The Methodology of Social Sciences*, eds, and trans. E. A. Shils and H. A. Finch (Glencoe, Ill., 1949), especially pp. 60, 57, 82ff.

14. Max Weber, 'Wissenschaft als Beruf' (1917), *Gesammelte Aufsätze zur Wissenschaftslehre*, pp. 588f, 601, 604f, translated as 'Science as a Vocation', in P. Lassman and I. Velody, eds, *Max Weber's Science as Vocation* (London, 1987), pp. 9, 20, 23. For an elegant illustration of the importance of these themes to understanding Weber as a whole, see Larry Scaff, *Fleeing the Iron Cage* (Berkeley, Calif., 1991), in English; and in German, Wilhelm Hennis, *Max Webers Wissenschaft vom Menschen* (Tübingen, 1996), although this is now available in English translation as *Max Weber's Science of Man*, trans. Keith Tribe (Newbury, 2000). I have attempted to build on these and other arguments in *The State of the Political: conceptions of politics and the state in the thought of Max Weber, Carl Schmitt, and Franz Neumann* (Oxford, 2003), especially 66ff.

15. A discussion can be found in Joseph Femia, *Against the Masses* (Oxford, 2001).

16. For a recent evaluation of the relationship between Max Weber and Robert Michels viewed through the lens of Michels's writings on patriotism and nationalism, see Duncan Kelly, 'From Moralism to Modernism: Robert Michels on the history, theory and sociology of patriotism', *History of European Ideas* 29 (2003), 339–63.

17. See especially Quentin Skinner, 'The Empirical and the Critical Theorists of Democracy: a plague on both their houses', *Political Theory* 1 (1972), 287–306.

18. Peter Laslett, 'Introduction', *Philosophy, Politics and Society*, 1st series, ed. Peter Laslett (Oxford, 1956); Isaiah Berlin, 'Does Political Theory Still Exist', in P. Laslett and W. G. Runciman, eds, *Philosophy, Politics and Society*, 2nd series (Oxford, 1964 [orig. 1962]), pp. 1–33, especially p. 28; David Easton, 'The Decline of Modern Political Theory', *Journal of Politics* 13 (1951), 36–58. This standard presentation of political science was not without challenge. For a notable paper, see Alasdair Macintyre, 'Is a Science of Comparative Politics Possible?' in A. Macintyre, *Against the Self-Images of the Age* (London, 1971), pp. 260–79.

19. Peter Novick, *That Noble Dream: the 'objectivity question' in North American historical scholarship* (Cambridge, 1988), pp. 597ff, 57–60; the quotation is from p. 628.

20. Richard Tuck, 'History of Political Thought', in P. Burke, ed., *New Perspectives on Historical Writing* (Oxford, 1995), pp. 193–205, especially pp. 197–201; John Rawls, 'Kantian Constructivism in Moral Theory', reprinted in *John Rawls: Collected Papers*, ed. S. Freeman (Cambridge, Mass., 1999), pp. 303–59, outlines again the philosophical presuppositions of his approach to moral theorising that informed the projects of *A Theory of Justice* (Cambridge, Mass., 1971), and would inform the later *Political Liberalism* (New York, 1993) and *The Law of Peoples* (Cambridge, Mass., 1999).

21. See R. G. Collingwood, *The Idea of History* (Oxford, 1956), pp. 215, 317.

22. For elegant rebuttals of these criticisms, see Quentin Skinner, 'The Principles and Practice of Opposition: the case of Bolingbroke versus Walpole', in N. McKendrick, ed., *Historical Perspectives* (London, 1974), pp. 93–128; 'Sir Geoffrey Elton and the Practice of History', *Transactions of the Royal Historical Society* (6th series) 7 (1997), 301–16.

23. Quentin Skinner, ed., *The Return of Grand Theory in the Human Sciences* (Cambridge, 1990).

24. See T. E. Willey, *Back to Kant: the revival of Kantianism in German social and historical thought, 1860–1914* (Detroit, Mich., 1978).

25. Wilhelm Dilthey, 'The Nature of Philosophy' and 'Construction of the Historical World' in *Dilthey: Selected Writings*, ed., H. P. Rickman (Cambridge, 1976), pp. 127, 177–84; cf. Hans-Georg Gadamer, *Truth and Method*, trans. W. Glyn (London, 1975), p. 150.

26. See Klaus Reich, 'Kant and Greek Ethics (I.)', *Mind* 48 (1939), 338–54; 'Kant and Greek Ethics (II.)' *Mind* 48 (1939), 446–63.

27. See the informative discussion in Larry Frohman, 'Neo-Stoicism and the Transition to Modernity in Wilhelm Dilthey's Philosophy of History', *Journal of the History of Ideas* 56 (1995), 263–87, especially 266, and the fascinating study that begins with Dilthey and studies the complex relationship between Nazism and the study of neo-Stoicism by Peter N. Miller, 'Nazis and Neo-Stoics: Otto Brunner and Gerhard Oestreich before and after the Second World War', *Past and Present* 176 (2002), 144–86, especially 182f.

28. Miller, 'Nazis and Neo-Stoics', p. 147; cf. Jonathan Israel, *Radical Enlightenment: philosophy and the making of modernity 1650–1750* (Oxford, 2000).

29. For the more broad-ranging assessment of Lipsian stoicism *pace* Oestreich's interpretation, see Arnaldo Momigliano, 'The First Political Commentary on Tacitus', *Journal of Roman Studies* 37 (1947), 91–101, especially 97f; 'Juste Lipsius et les Annales de Tacite. Une Methode de Critique Textuelle au XVIe Siecle', *Journal of Roman Studies* 39 (1949), 190–2, especially 192; this provides the basis for certain themes developed by Richard Tuck, *Philosophy and Government, 1572–1651* (Cambridge, 1993), pp. 45–64.

30. See Miller, 'Nazis and Neo-Stoics', especially 155–8 (on Brunner and the rise of *Ostforschung* in particular); 161–3, 173–8 (on Oestreich's militarist version of Lipsian neo-Stoicism, especially p. 162, n. 50, n. 51, where he notes that the editors of the celebrated English translation of Gerhard Oestreich, *Neo-Stoicism and the Modern State*, eds, B. Oestreich and H. G. Koenigsberger, trans. D. McLintock (Cambridge, 1982), expunge the dedicatory notes to Brunner that were present in various of the original articles that make up that translation); 164ff, 172 (on Oestreich's debts to Otto Hintze). See also Carlo Ginzburg, 'Germanic Mythology and Nazism: thoughts on an old book by Georges Dumézil', *Myths, Emblems, Clues*, trans. Anne E. Tedeschi (London, 1990), pp. 126–45, noting the scholarly connections between Dumézil and Marc Bloch in particular. Cf. Arnaldo Momigliano, 'Georges Dumézil and the Trifunctional Approach to Roman Civilization', *History and Theory* 22 (1984), 312–330, especially 329 on the foundational difference between a 'trifunctional' Christian order, and a non-trifunctional Pagan one, that renders it difficult to see Indo-European trifunctionality in Dumézil's interpretation of Roman history. For more recent scholarship on the relationship between German historians and National Socialism, see, for example, Peter Schöttler, ed., *Geschichtsschreibung als Legitimationswissenschaft, 1918–1945* (Frankfurt, 1997); Ingo Haar, *Historiker im Nationalsozialismus* (Göttingen, 2000).

31. Miller, 'Nazis and Neo-Stoics', 158, 183–6.
32. Marc Bloch, 'Pour une histoire comparée des sociétés européenes', *Mélanges Historiques*, 2 Vols (Paris, 1963), I, pp. 16–40; *Les rois thaumaturges* (Strasbourg, 1924), translated as *The Royal Touch: sacred monarchy and scrofula in England and France*, trans. J. E. Anderson (London, 1973).
33. See Lucien Febvre, *La terre et l'évolution humaine* (Paris, 1922); Fernand Braudel, *The Mediterranean and the Mediterranean World in the Age of Phillip II*, trans. Siân Reynolds, 2 Vols (London, 1972); 'History and the Social Sciences: the *longue durée*' in *On History*, trans. Sara Matthews (Chicago, Ill., 1980), pp. 25–54.
34. See the discussion in Perry Anderson, 'Fernand Braudel and National Identity', *A Zone of Engagement* (London, 1992), p. 265, n. 40; for a different analysis, stressing the overall importance of Elie Halévy in particular, see Julia Stapleton, *Englishness and the Study of Politics: the social and political thought of Ernest Barker* (Cambridge, 1994), especially pp. 120–7.
35. For an engaging account of the 'School', see Peter Burke, *The French Historical Revolution: the Annales School, 1929–1989* (Oxford, 1990).
36. François Furet, *Interpreting the French Revolution*, trans. Elborg Foster (Cambridge, 1981); Robert Darnton, *The Business of Enlightenment* (Cambridge, Mass., 1979); Roger Chartier, *The Cultural Origins of the French Revolution*, trans. Lydia G. Cochrane (London, and Durham, NC, 1991); Pierre Nora, ed., *Realms of Memory*, 3 Vols, ed. L. D. Kritzman (New York, 1996).
37. See especially Keith Baker, *Inventing the French Revolution* (Cambridge, 1990).
38. Keith Baker, 'In Memoriam: François Furet', *Journal of Modern History* 72 (2000), 1–5; Michael Scott Christofferson, 'An Antitotalitarian History of the French Revolution: François Furet's *Penser la Revolution française* in the intellectual politics of the late 1970s', *French Historical Studies* 22 (1999), 557–611, and the earlier Forum on 'François Furet's Interpretation of the French Revolution', *French Historical Studies* 13 (1990), 766–802.
39. See also François Furet, 'Beyond the Annales', *Journal of Modern History* 55 (1983), 389–410.
40. François Furet, *The Passing of an Illusion* (Chicago, Ill., 1999).
41. For celebrated discussion, see the essays by Raphael Samuel, 'British Marxist Historians Part I', *New Left Review* 120 (1980), 21–96; also Eric Hobsbawm, 'The Historians' Group of the Communist Party', in M. Cornforth, ed., *Rebels and Their Causes: essays in honour of A. L. Morton* (London, 1978), pp. 21–47, and Perry Anderson, *Arguments Within English Marxism* (London, 1980).
42. For two illustrations only, see Neal Wood, 'The Social History of Political Theory', *Political Theory* 6 (1978), 345–67, especially pp. 345, 351, 361, 364; Ellen Meiksins Wood and Neal Wood, *Class Ideology and Ancient Political Theory: Socrates, Plato and Aristotle in social context* (Oxford, 1978), pp. ix, 1, 12.
43. See Robert Darnton, 'In Search of the Enlightenment: recent attempts to create a social history of ideas', *Journal of Modern History* 43 (1971), 113–32 (132); 'Intellectual and Cultural History', *The Kiss of Lamourette: reflections in cultural history* (New York, 1990), pp. 191–218.
44. See Duncan Kelly, 'The Political Thought of Isaiah Berlin', *British Journal of Politics and International Relations* 4 (2002), 25–48, especially 43ff.
45. But for the context of the rise of the study of political thought in English universities, see the intriguing discussion by Robert Wokler, 'The Professoriate of Political Thought in England since 1914: a tale of three chairs', in D. Castiglione and I. Hampsher-Monk, eds, *The History of Political Thought in National Contexts* (Cambridge, 2001), pp. 134–58.

46. See John Plamenatz, *Man and Society*, 3 Vols (London, 1963), and G. H. Sabine's famous textbook, *A History of Political Theory* (3rd edn, London, 1963).

47. See C. B. Macpherson, *The Political Theory of Possessive Individualism* (Oxford, 1962); this was a work as much about its own moment, as illustrated by Macpherson's discussion of the possibilities of nuclear annihilation, as it was about the context of Hobbes and Locke. The thesis was subject to criticism too numerous to mention, but a fair-minded defence of Macpherson that details most of these criticisms can be found in Jules Townshend, *C. B. Macpherson and the Problem of Liberal Democracy* (Edinburgh, 2000).

48. Thomas Kuhn, *The Structure of Scientific Revolutions* (2nd edn, Chicago, Ill., 1970), especially pp. 166ff.

49. Among many profoundly important publications, Momigliano's Sather Lectures on *The Classical Foundations of Modern Historiography* (Berkeley, Calif., 1990) perhaps stand out for their impact. His own account of the modern study of the history of ideas can be seen in the pithy essay 'A Piedmontese View of the History of Ideas', *Essays in Ancient and Modern Historiography* (Oxford, 1977), pp. 1–9, especially p. 5, noting the important work of Venturi though stressing in fact his 'French' style to account for his impact; on p. 6 he suggests that the history of ideas has most to offer when it studies distant and indeed perhaps alien cultures; and on p. 4, he notes that 'the price English historians paid in the 1930s for remaining independent of German *Ideengeschichte* was to jettison their own tradition of the history of ideas'. See also Franco Venturi, *Utopia and Reform in the Enlightenment* (Cambridge, 1971); *The End of the Old Regime in Europe*, 2 Vols (Princeton, NJ, 1991); *Italy and the Enlightenment* (London, 1972). It is no coincidence that the first two volumes of J. G. A. Pocock, *Barbarism and Religion* (Cambridge, 1999–) are dedicated to Venturi and Momigliano respectively. Momigliano's legacy has been developed further and in slightly different directions by the equally voluminous scholarship of Anthony Grafton in particular. See his *Bring Out Your Dead: the past as revelation* (Cambridge, Mass., 2004) for one illustration.

50. Carlo Ginzburg, 'Microhistory: two or three things I know about it', trans. John and Anne C. Tedeschi, *Critical Inquiry* 20 (1993), 10–35 (17, 20).

51. Geoff Eley, 'Labor History, Social History, *Alltagsgeschichte*: experience, culture, and the politics of the everyday – a new direction for German social history?', *Journal of Modern History* 61 (1989), 297–343; cf. G. Iggers, ed., *The Social History of Politics: critical perspectives in West German historical writing since 1945* (Leamington Spa, 1985), for earlier illustrations.

52. For the wider attempt to retain a specifically 'intellectual' focus for intellectual history, see Annabel Brett, 'What is Intellectual History Now?' in D. Cannadine, ed., *What is History Now?* (Basingstoke, 2002), pp. 113–31 (p. 127); itself an attempt to rebut some of the claims made about intellectual history and Ginzburg's own methods by Roger Chartier, 'Intellectual History or Sociocultural History?' in Dominick LaCapra and Steven L. Kaplan, eds, *Modern European Intellectual History: re-appraisals and new perspectives* (Ithaca, NY, 1982), pp. 13–46; cf. Dominick LaCapra, *History and Criticism* (Ithaca, NY, 1985), especially pp. 79–94 on the importance of 'culture' to the history of *mentalités*. A recent edition of the leading journal of intellectual history has taken as its theme the contemporary question of 'Intellectual History in a Global Age', *Journal of the History of Ideas* 66 (2005), in an attempt to locate traditional questions of intellectual history within a contemporary compass.

53. See Carlo Ginzburg, '"Your Country Needs You": A Case Study in Political Iconography', *History Workshop Journal* 52 (2001), 1–22; more generally, see his

'From Aby Warburg to E. H. Gombrich: a problem of method' in *Myths, Emblems, Clues*, especially pp. 20f. There is a very attractive critical discussion of Ginzburg and his approach by Perry Anderson, 'Nocturnal Enquiry: Carlo Ginzburg' in *A Zone of Engagement* (London, 1992), pp. 207–29.

54. See Duncan Kelly, 'Rethinking Franz Neumann's Route to *Behemoth*', *History of Political Thought* 23 (2003), 458–96.

55. See Franz Neumann, *The Rule of Law* (Leamington Spa, 1986 [orig. 1936]); *Behemoth: The Structure and Process of National Socialism, 1933–1944* (New York, 1944). There are intriguing brief comments by another great émigré scholar of eminence about his relationship with Neumann as a senior colleague and which give a flavour of Neumann's importance, in Peter Gay, *My German Question* (New Haven, Conn., 1998), pp. 2ff.

56. For part of the story on the development of North American political science, see John Gunnell, *Imagining the American Polity: political science and the discourse of democracy* (Pennsylvania, 2005). Franz Neumann, *The Democratic and Authoritarian State*, ed., Herbert Marcuse (New York, 1957), contains selections of his post-war writings that highlight this concern with the history of political thought and the rule of law.

57. An interesting Straussian reflection by one of his most famous pupils can be found in Allan Bloom, 'Leo Strauss: September 20, 1899–October 18, 1973', *Political Theory* 2 (1974), 372–92, especially 383f on the 'discovery of esoteric writing'.

58. See Jürgen Habermas, 'Constitutional Democracy: a paradoxical union of contradictory principles', *Political Theory* 29 (2001), 766–81; *The Inclusion of the Other*, eds C. Cronin and P. de Grieff (Oxford, 1998); cf. Patchen Markell, 'Making Affect Safe for Democracy? On "constitutional patriotism"', *Political Theory* 28 (2001), 38–63.

59. Jürgen Habermas, 'The Horrors of Autonomy: Carl Schmitt in English', *The New Conservatism: cultural criticism and the historians debate*, ed. and trans. Shierry Weber Nicholsen (Oxford, 1994), pp. 128–39.

60. See Jan-Werner Müller, *A Dangerous Mind: Carl Schmitt in post-war European thought* (London, and New Haven, Conn., 2004), and the very brief discussion of the hermeneutical analyses of *Begriffsgeschichte* and its place in the history of intellectual history in Donald R. Kelley, *The Descent of Ideas* (Aldershot, 2002), pp. 235ff. Reinhart Koselleck's most recently available discussion in English on these themes is *The Practice of Conceptual History: timing history, spacing concepts*, trans. Todd Presner, Kerstin Behnke and Jobst Welge (Stanford, Calif., 2002); a presentation of his disagreements with Gadamer can also be found in *Zeitschichten* (Frankfurt, 2000).

61. For the wider context of post-war French political thought, see, for example, Sunhil Khilnani, *Arguing Revolution: the intellectual left in post-war France* (New Haven, Conn., 1993); Tony Judt, *Marxism and the French Left: essays on labour and politics in France, 1830–1981* (Oxford, 1989).

62. For Foucault's development and reading of Oestreich, see Pasquale Pasquino, 'Michel Foucault (1926–1984): the will to knowledge', *Economy and Society* 15 (1986), 97–109. Pasquino's utilisation in turn of Foucault's arguments about discipline as well as governmentality informed his own impressive concerns with the languages of *Cameralwissenschaft* and the relationship between government and happiness in the discourses of 'police' in eighteenth-century Europe in general, and Germany in particular. See his '*Theatrum Politicum*: genealogy and the science of police', *Ideology and Consciousness* 4 (1978), 41–54.

63. This point is well made by Daniel Conway, *'Pas de Deux*: Habermas and Foucault in genealogical communication', in S. Ashenden and D. Owen, eds, *Foucault contra Habermas* (London, 1999), pp. 60–89, especially pp. 62, 69–73, 75–79; and among many possible citations, Michel Foucault, 'What is Enlightenment?' *Ethics: Subjectivity and Truth*, ed., Paul Rabinow, trans. Robert Hurley et al. (London, 1997), pp. 303–20.

64. Jürgen Habermas, *The Philosophical Discourse of Modernity*, trans. Frederick Lawrence (Oxford, 1995 [orig. 1985]), pp. 256ff, 264f.

65. For a sensible contribution, see Charles Maier, *The Unmasterable Past* (Cambridge, 1988). The brilliant essay by Geoff Eley, 'Nazism, Politics and the Image of the Past: thoughts on the West German *Historikerstreit* 1986–1987', *Past and Present* 121 (1988), 171–208, especially 185, 202–7, and *passim*, should also be consulted.

66. Jürgen Habermas, 'Eine Art Schadensabwicklung: Die apologetischen Tendenzen in der deutschen Zeitgeschichtsschreibung', *Die Zeit* (11 July 1986), was the foundational intervention, and is reprinted in full, without the editing undertaken on the original piece, in English translation, as 'Apologetic Tendencies', in Habermas, *The New Conservatism*, pp. 212–28.

67. Cf. Hayden White, *Metahistory: the historical imagination in nineteenth-century Europe* (Baltimore, Md, 1973); Lawrence Stone, 'The Revival of Narrative', *Past and Present* 85 (1979), 3–24.

68. See Martin Jay, 'Should Intellectual History Take a Linguistic Turn? Reflections on the Habermas–Gadamer debate', in LaCapra and Kaplan, *Modern European Intellectual History*, pp. 86–110, pp. 106f.

69. See Bernard Williams, *Truth and Truthfulness* (Princeton, NJ, 2003), which focuses on the relationship of truth to questions of accuracy and sincerity, which must be understood historically. The appropriate comparison, perhaps, is with Richard Rorty, 'The Historiography of Philosophy: four genres', in R. Rorty, J. B. Schneewind and Q. Skinner, eds, *Philosophy in History* (Cambridge, 1993 [orig. 1984]), pp. 49–75, as well as with *Objectivity, Relativism, and Truth: philosophical papers* (Cambridge, 1991).

70. Ernst Gombrich, *Aby Warburg – an intellectual biography* (London, 1970), pp. 287, 308, 327; for a slightly different and more recent reflection by the late Director of the Warburg Institute, see E. H. Gombrich, 'Aby Warburg: his aims and methods: an anniversary lecture', *Journal of the Warburg and Courtauld Institutes* 62 (1999), 268–82.

71. The journal was initially founded as the *Journal of the Warburg Institute* in 1937, and renamed in 1939.

72. Nicolai Rubinstein, 'The Beginnings of Political Thought in Florence: a study in medieval historiography', *Journal of the Warburg and Courtauld Institutes* 5 (1942), 198–227; 'Political Ideas in Sienese Art: the frescoes by Ambrogio Lorenzetti and Taddeo di Bartolo in the Palazzo Pubblico', *Journal of the Warburg and Courtauld Institutes* 21 (1958), 179–207. Such themes have been taken up by many, including Quentin Skinner, 'Ambrogio Lorenzetti: the artist as political philosopher', *Proceedings of the British Academy* 72 (1987), 1–56; 'Ambrogio Lorenzetti's *Buon governo* Frescoes: two old questions, two new answers', *Journal of the Warburg and Courtauld Institutes* 62 (1999), 1–28.

73. Ernst Kantorowicz, 'The Este Portrait by Roger van der Weyden', *Journal of the Warburg and Courtauld Institutes* 3 (1940), 165–80; 'Ivories and Litanies', *Journal of the Warburg and Courtauld Institutes* 5 (1942), 56–81, this latter article laying some of the foundations for his later study *Laudes Regiae: a study in liturgical acclamations*

and mediaeval ruler worship (Berkeley, Calif., 1946). Whilst Kantorowicz is most well known for his study *The King's Two Bodies* (Princeton, NJ, 1997 [orig. 1957]), his earliest work, *Kaiser Friedrich der Zweite* (Berlin, 1927), has attracted numerous critical commentaries for its politics. For a wide-ranging assessment, see Martin Ruehl, '"In this Time Without Emperors": the politics of Ernst Kantorowicz's *Kaiser Friedrich der Zweite* reconsidered', *Journal of the Warburg and Courtauld Institutes* 63 (2000), 187–242, especially 217–20.

74. See Ginzburg, 'From Aby Warburg to E. H. Gombrich', pp. 23ff; Fritz Saxl and Erwin Panofsky, 'Classical Mythology in Mediaeval Art', *Metropolitan Museum Studies* 4 (1932/33), 228–80, was the early study.

75. See Francis Haskell, *History and its Images* (London, and New Haven, Conn., 1995), who illustrates this use of images by historians.

index